Light Vehicle Body Repair and Refinishing

Light Vehicle Body Repair and Refinishing

Frank Andrews

Vehicle Series Editor: Roy Brooks

MACMILLAN

First published 1996 by
MACMILLAN PRESS LTD
Houndmills, Basingstoke, Hampshire RG21 6XS
and London
Companies and representatives
throughout the world

ISBN 0–333–63398–9

A catalogue record for this book is available
from the British Library.

10   9   8   7   6   5   4   3   2   1
05   04   03   02   01   00   99   98   97   96

Copy-edited and typeset by
Povey–Edmondson
Okehampton and Rochdale, England

Printed in Great Britain by Unwin Brothers Ltd
The Gresham Press, Old Woking, Surrey
A member of the Martins Printing Group

To the late
Ted Cooper,
one time Paint and Body Manager, VAG United Kingdom Ltd,
and his successor,
Andy Sargent.
Without these gentle and most patient tutors
this book could never have been written

Contents

List of Figures

List of Plates

Preface

This book has been written to provide the underpinning knowledge that will be needed by those hoping to qualify in light vehicle body repair and refinishing, an industry that is changing dramatically. Developments which have been progressing at a more leisurely industrial pace have been accelerated by the demands for a better environment, both inside the workshop and in the world outside. The unrelenting pressure of the insurers to cut costs, while at the same time reinstating vehicles which are more exacting to repair and refinish, has brought a need for the highest standards of management and of technical competence within the bodyshop.

In this book is gathered the wisdom and knowledge of many in the forefront of the industry: specialists from the vehicle manufacturers, makers and suppliers of paint and materials, equipment manufacturers and those who teach and train. Above all, those who work within the industry have contributed an essential practicality. In that respect this book goes beyond providing underpinning knowledge in that it seeks to set out a consensus view of what is currently the best practice in light vehicle body repair and refinish.

If I have failed in any way, it is from a desire to take the best from all the interests involved rather than to set out any one view above another.

FRANK ANDREWS

Acknowledgements

The principal advisers, suppliers of information and material for reproduction in this book are members of the Vehicle Makers Body Repair Council. My deepest thanks are due to the chairman, Mike Sear of Jaguar, and to those members of the council who have provided these foundations. In particular, I am indebted to David Lester, Bodyshop Development Manager of Porsche Cars Great Britain Ltd, who volunteered – without prompting – to read the entire draft typescript!

The paint sections have been the subject of close scrutiny by Kevin McDermott, Technical Training Manager of Glasurit Automotive Refinish, and Larry Fernandes, Standox Technical Manager, who have made sure that these chapters are as accurate as it is possible to be at this time. ICI Autocolor kindly permitted the use of the gun setting extract from 'The Refinishers Handbook' in Chapter 7, which they amended to suit present day conditions.

I am indebted to Bob Driver of Minden Industrial, Ryan Holmes of 3M® (United Kingdom) PLC, Dennis Relph of Globeaid, Ken Richardson of the Welding Institute, Paul Smith of Celette UK Limited, John and Michael Stanners of Stanners Paints Ltd, and many others who sent material or answered questions. My special thanks are due to Roy Brooks and others from education who painstakingly read every word and provided much constructive comment.

Many other companies have also contributed and I acknowledge and thank: Beta Industrial Ltd, Binks-Bullows Ltd, Blackhawk Automotive Ltd, Robert Bosch GMBH, Celette UK Ltd, Devilbiss Automotive Refinishing Products, Elcometer Instruments Ltd, C and E Fein Gmbh, Ford Motor Co Ltd, Fordley Laminated Notices Ltd, Glasurit Automotive Refinish, Herbert's Standox, ICI Autocolor, IRT Infrarödteknik AB, Sweden, Jaguar Cars Ltd, Moldrex-Metric AG and Co KG, The Motor Insurance Repair Research Centre (Thatcham), Peugeot Talbot Motor Co Ltd, 'Bodyshop Magazine and Bodyshop Buyers Guide' – Plenham Ltd, Porsche Cars Great Britain Ltd, Saab Great Britain Ltd, Sheen Instruments Ltd, Unic International (UK) Ltd, VAG United Kingdom Ltd, Verlagsgruppe Bertelsmann International GmbH and Welwyn Tool Co Ltd.

I can confirm that writing a book is not possible without the forbearance, support and practical help of a partner and to my wife, Mary, I owe especial thanks. Her skills in entering information into the computer, checking my drafts and providing a constant supply of refreshment were quite invaluable.

FRANK ANDREWS

Trade Marks and Other References

Due acknowledgement is made for the use of the following trade marks and titles:

Chapter 3

TORX® is a registered trade mark of Camcar division of Textron Inc.

Chapter 4

Porto-Power® is a registered trade mark of Blackhawk® International.

Chapter 5

'Reversing the Accident' – acknowledgements to Blackhawk Automotive Limited.
Dozer® is a registered trade mark of Blackhawk® International.
Scotchbrite® is a registered trade mark of 3M® United Kingdom PLC.

Checklist of NVQ Units, Chapters and Sections

Glossary

A pillar	The first pillar supporting the roof adjacent to the windscreen
Activator/Hardener	A chemical which will cause a curing process, the highly toxic ingredient of two–pack paints
Active safety	Features that help to prevent an accident happening
Additive mixing	The mixing of the colours of light
Adhesion	Joining together of two components, as of paint to a substrate or one component to another
Air-fed mask	Painter's breathing equipment with an independent air supply
Anti-corrosion	A treatment or construction that prevents or delays the formation of rust on ferrous metal
Arc eye	The effect of an arc welding flash on the eye
Automated	An action that is carried out by machinery without human intervention
B pillar	The second pillar back from the screen
Blending-out	A spraying action where newly applied paint is encouraged to blend with the existing finish
Booth	An enclosed room with filtered air flow in which a car can be spray painted, usually now has heating when it is called a booth/oven
BS	British Standard
Burnishing	Polishing, generally using a gentle abrasive compound
Cellulose	Sometimes called nitrocellulose, it is a highly flammable, quick drying paint
Clinching	Joining materials by pressing two or more layers together into a locking shape
Cohesion	The force by which particles in a substance are held together
Colour library	Painter's record of bodyshop formulated colours
Consumables	Materials which are consumed, lost or destroyed by being used, e.g. sanding discs, masking tape, etc.
Contamination	The inclusion of unwanted substances in or under a paint film
Contour	Shape or outline
COSSH	Control of Substances Hazardous to Health Regulations 1988
Daylight lighting	Artificial lighting that is compensated to produce the same effect as daylight
De-mineralised water	Water from which mineral particles of all but the smallest size have been removed
De-nibbing	The action of removing small dirt inclusions or other imperfections on a coat of paint
Dirt inclusions	Small particles of dust or fluff that have become embedded in the surface during painting
Dry sanding	Literally, sanding bodywork without using water
Dust extraction	The removal of dust from a sanding tool by means of vacuum, often from a central extraction unit
Edge-to-edge finish	Painting a panel or panels up to gaps, joins or trim so that no blending is necessary

Electromagnetic spectrum	The radiation given off by the sun; the wavelengths range from Gamma rays (about a billionth of a millimetre long) to Radio waves (which can be a kilometre long)
Electrophoresis	Also described as cataphoresis, particles of paint are attracted to a submerged car body by a flow of electric current
Electrophoretic primer	Primer applied by electrophoresis, as with the E coat on a car body or component
Electrostatic process	The attraction of paint particles to an electrically charged car body
EN	Standards that are being adopted in all the states of the European Union
EP	Epoxide; Epoxy resin
EPA	Environmental Protection Act
Etching	An acidic action that cuts into the surface to improve adhesion
Feathering	Relating to paint preparation, it means sanding away all edges to provide a smooth, even surface
FEPA	Federation of European Producers of Abrasives
Filler	A material that is applied to make good surface irregularities
Flashing-off	The partial drying of paint to the point where it loses its wet gloss and looks matt
Flatted	Taking away the gloss finish of the paint with an abrasive
Formulation	The 'recipe' which gives the mix of bases and tinters for making up a paint colour
Galvanised	Metal, usually steel, that is coated with a thin layer of zinc by either a hot dip or electrolytic process
GRP	Glass Reinforced Plastic
Guidecoat	A contrasting colour applied to primer filler to show material removal during sanding
HSS	High Strength Steel
HVLP	High Volume, Low Pressure
Immersed	Lowered beneath the surface, such as a car body in a tank of paint
Infra-Red	A type of heater used to cure paints and fillers
Ingestion	Taking in substances through the mouth
Isocyanates	A highly toxic ingredient of the activator for two-pack (2K) paints
Key	The removal of gloss from a surface using a fine abrasive so that other coatings can adhere to it
Lacquer/Clear Coat	A clear paint applied over the colour coat to give gloss and durability
MAGS	Metal Arc Gas Shielded welding
Masked	The temporary attachment of a protective material such as tape, shaped foam, sheeting or a film to protect a surface
Metallic and pearlescent basecoats	Basecoats which contain particles to give special effects, luminium for metallic and mica for pearlescent
Metallic finishes	Paints containing aluminium particles that reflect light
Metamerism	The name used to describe the phenomenon where colour changes shade under different lighting, e.g. sodium vapour street lights
Microfiche	Transparent film on which is printed minute lettering or illustrations which can be read under a viewer
Micron	A millionth of a metre or one thousandth of a millimetre
MIG/MAG	Metal Inert Gas/Metal Active Gas arc welding
MIRRC	Motor Insurance Repair Research Centre (Thatcham)
Mixing Scheme	A motorised rack that holds cans of paint and stirs them
Mule skinner	Rotary wire brush which has its wires embedded in resin
OE	Original equipment (of a car, fitted or applied at the factory)

Oven	A heated painting booth
Overbaked	Relating to car production lines, this happens when the moving conveyor slows down or stops and the car body spends too long in the oven, resulting in excessively hard paint
PA	Polyamide
Paint booth	An enclosed and vented cubicle in which paint can be sprayed without contaminating the rest of the work area
Passive safety	Those measures that minimise the consequences of an accident
PBT	Polybutylene terephalate (linear Polyester)
Peroxide bleaching	A stain which passes through the topcoat from beneath, caused by poor measuring or mixing of two–pack fillers
PC	Polycarbonate
PE	Polyethylene
PES	Polyethersulphone
PET	Polyethylene terephthalate
PF	Phenol-formaldehyde resin
Phosphate	A paint containing zinc particles in sufficient quantity to delay or prevent rust formation
Phosphatising	The process for applying a phosphate coating
Plasticiser	An additive that gives a paint film the flexibility of plastic
PMMA	Clear Acrylic or Poly (methyl methacrylate)
POM	Polyoxymethylene; polyformaldehyde
Pot-life	The time during which mixed adhesive, filler and paint can be used
PP	Polypropylene
PPE	Personal protective equipment
PPE	Polyphenylene ether
PPO	Modified polyphenylene oxide
PPS	Polyphenylene sulphide
Primary colours	Those of light are red, blue and green which together make white, pigments are magenta (a red), cyan (a blue) and yellow which, when mixed together, make black
PTFE	Polytetrafluoroethylene
PUR	Polyurethane
PVC	Polyvinyl chloride
Random orbital	Sanders so described have a plate or disc which moves in a constantly varying orbit, those with a round head also rotate and are known as dual action or DA sanders
Roll-edge masking	Masking which presents a gradually diminishing gap to paint spray which can help in avoiding sharp edges
Sensitised	Heightened sensitivity – in the case of isocyanates the slightest trace can provoke a reaction
SMA	Styrene maleic anhydride
Sodium azide	A highly toxic constituent of airbag gas cartridges
Solid colours	Paints which rely only on coloured pigments for effect
Spray-out card	A special card that is painted with the mixed paint for comparison or a permanent record
Stearate powder	A dry lubricant applied by the makers to dry sanding abrasive
Stopper	Usually a one-pack material that cures by evaporation
Substrate	The material that is painted
Subtractive mixing	The mixing of pigment colours
Suspension	Relating to paints it refers to particles mixed in the paint and how well they are mixed
Synthetic paints	Paints which contain a resin
Tacking-off	Wiping dust from a surface to be painted with a dust attracting cloth
TAGS	Tungsten Arc Gas Shielded welding

Thinner	A solvent that reduces the viscosity of paint so that it may be sprayed and evaporates as the paint dries.
TIG	Tungsten Inert Gas arc welding
Topcoat	The outer finishing coat; basecoat and clear finish may also be considered as topcoat
TPUR	Thermoplastic polyurethane
Translucent coating	A cloudy coat that permits some light rays to pass through to the layer below
Transparent coating	A clear coat that permits light rays to pass through to the layer below
Two-pack acrylic	A resin based paint that hardens chemically with the addition of a hardener, usually written 2K and called 2-pack
Viscosity cup	A cup of measured size which also has a hole of measured size in the base to check paint viscosity by time
VOCs	Volatile Organic Substances
Volatile	Capable of readily evaporating

Chapter 1

Health and Safety at Work

Introduction

This book begins and ends with the subject of people. Customers are, perhaps, the most important, for without them you would have no work. Your workmates, too, are important, for no business can be truly successful unless its workers make an efficient team. But you are the main subject of this first chapter.

You probably spend at least a third of your waking hours at work, so it is right that we should start by considering how well you look after yourself in the workplace. If you want to enjoy your spare time you must spend those working hours wisely. When you are working hard under pressure you must also work safely, so that you can enjoy good health, and the money that you earn. In this first chapter we are going to look at what is involved in a general way. More specific guidelines for each type of work are given throughout the book.

You probably already know that from the moment you arrive on your college or company premises until the moment you leave, your actions are governed by many regulations. Those that cover bodyshops are amongst the most diverse and comprehensive of any section of industry. You probably also know that they can be irksome, often to the point of discomfort. But they are there to protect **you**, sometimes from yourself!

Your health and safety at work

The working environment

The bodyshop has a great number and a wide variety of potential hazards (see Fig. 1). Here are some of them:

Figure 1 These are reminders – not decorations, do what the safety signs say – and stay healthy

- There are dangers that can be seen clearly: obstructions like cables and hoses, components and equipment that can cause you to trip over them.
- The much less noticeable and often invisible chemical hazards, such as the fumes from solvents, which can damage your body's internal systems.
- Poisons abound, along with compressed air and gases.
- Noise can permanently impair your hearing and the flash of welding hurt your eyes.
- Dust and spray mist can block the airways of your lungs.

You can stay safe and healthy if you and your team-mates look after yourselves and care for each other.

Attending the result of traffic accidents at the roadside is another potentially dangerous environment in which you may have to work. Fortunately, before you arrive the emergency services should have dealt with any drivers and passengers who have injuries. Your involvement will be to get on with the job of loading a damaged vehicle for removal to your bodyshop. The weather and the close proximity of moving traffic may make work difficult. Care must still be exercised in handling the vehicle if additional, expensive damage is to be avoided.

In what are sometimes difficult circumstances, you must be aware that it is all too easy to pull muscles or sustain cuts and abrasions. Inactivated airbags may also call for special care to avoid their inadvertent release or contact with any chemicals that may have spilt from them.

Sometimes a worried, perhaps shocked, owner or driver is there with the vehicle. Do remember that people are more important than any vehicle. (The handling of customers is considered in greater detail in Chapter 10.)

Regulations affecting activity

There are many regulations which a bodyshop must observe. Some relate to protecting the outside environment while others are concerned with what happens inside the bodyshop; they protect you at work. These are some of the principal regulations that affect day-to-day work in the bodyshop:

- *Factories Act 1961*. Since 1830, working conditions have been regulated by various Factory Acts. Some of the provisions of the act still apply but many have been replaced by the new COSSH regulations.
- *Offices, Shops and Railway Premises Act 1963*. Working conditions such as heat and light in any office, reception or parts shop are covered by this Act. It is a requirement of the later Health and Safety at Work Act 1974 so it certainly applies to your company.

- *Fire Precautions Act 1971.* Under these regulations the Fire Service became responsible for inspecting industrial premises. They issue the fire certificate to show that the necessary fire fighting equipment, fire escapes and access to them have been provided.
- *The Health and Safety at Work Act 1974.* This is called an 'enabling Act', which means that the Government can introduce regulations that come under the act without the approval of Parliament. The most widely known are the COSSH Regulations of 1988.
- *Control of Pollution (Special Waste) Regulations 1980.* All bodyshops and their workers are required to separate, handle and dispose of waste correctly.
- *Health and Safety (First Aid) Regulations 1981.* This regulation requires that approved first aid is available and that some personnel in the company are trained as 'first aiders'.
- *Reporting of Injuries, Diseases and Dangerous Occurrences Regulations 1985 (RIDDOR).* These regulations extend reporting to all workers, including the self-employed and trainees, and call for all but the most minor occurrences to be reported.
- *Control of Asbestos at Work Regulations 1987.* At one time, asbestos was a principal constituent of brake and clutch linings. The dust arising from wear of these components is harmful if inhaled. Gaskets may also contain asbestos. The regulations require protection from inhalation.
- *Control of Substances Hazardous to Health Regulations 1988 (COSSH).* These are the single most important set of regulations and they affect everyone working in a bodyshop (see Fig. 2). Much of the guidance in the remainder of this book is to help you comply with the requirements of COSSH.
- *Noise at Work Regulations 1989.* Even if you are used to loud noise in your spare time, when at work your employer is required to monitor and control the level of noise in your workplace. You, for your part, must wear the ear defenders that he is obliged, by law, to provide if the noise level is too high.
- *Electricity at Work Regulations 1989.* The safe use of electrical equipment is the part of these regulations that most directly affects you. Never use dangerous equipment or use it in a way that could be dangerous.
- *Pressure Systems and Transportable Gas Container Regulations 1989.* Less well known, these regulations ensure that the compressed air system to be found in every bodyshop is installed correctly and routinely examined and maintained. Never do any work on the compressed air system without your employer's express permission.
- *Environmental Protection Act 1990.* This is also an 'enabling Act' which allows the Government to introduce regulations such as those next in our list.
- *Environmental Protection (Prescribed Processes and Substances) Regulations 1991* and *Environmental Protection (Applications, Appeals and Registers) Regulations 1991.* You will be aware that much concern has, and is, being shown about damage caused to our environment, both locally and world-wide. The make-up of the air that we breathe can be altered by solvent vapours, such as those released by paint spraybooths. Because of this, body-

Figure 2 A typical safety wall sign; get to know them – your health and even your life may depend upon it!

shops now come under much stricter local authority control and must comply with the regulations as they are implemented. These are intended at the present time to reduce evaporating solvents down to very low levels.
- *Waste Management – The Duty of Care – A Code of Practice 1991.* This is another regulation made under the Environmental Protection Act 1990. It applies to everybody, everywhere on the mainland of Great Britain. Much of the waste from bodyshops is potentially dangerous and some of it is liable to self-ignition. It is as well to know that you can be prosecuted for failing in the duty of care without actually committing an act that results in harm or pollution!
- *Personal Protective Equipment (European Directive) Regulations 1992.* These are part of Europe-wide legislation to provide a common standard for personal protective equipment (PPE) throughout Europe. Equipment approval will be in the hands of the makers and Notified Bodies, who will issue a CE mark for the product (see Figs 3 and 4).

Your duty and how you benefit

It is perhaps unfortunate, but true, that complicated regulations are needed to make us look after ourselves and those around us. One reason is that much of our work doesn't – at the time – seem to hurt us. A solvent gives off an aroma but

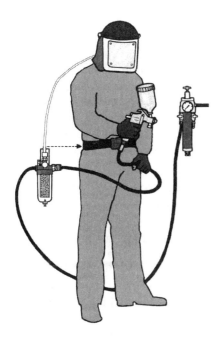

Figure 3   A high standard of health and safety protection for a painter:
filtered air from the regulator for both breathing and painting
an activated charcoal and colour indicated breathing air filter and regulator
silenced air outlets in the condensation-free visor
lint free overalls and solvent proof gloves
and non-sparking workboots or workshoes

Figure 4   A welder with adequate protection, the 3M® 06920 Welding
Fume Respirator provides protection against metal fume and ozone

we can manage without a mask. Or there's a little dust
around when we do a quick sanding job but it doesn't stop
us breathing. 'This cutting tool is making a lot of noise, but I
don't need ear defenders, it will only take a minute.' That
may all be true, but the poison builds little by little, the dust
clogs respiratory passages little by little and the noise is yet
more on top of the normal dose administered to your ear-
drums, some of it self-inflicted! Those people who have
been adversely affected by poor working practices might
well say 'Don't be a fool!' 'You can't rebuild good health like
a damaged motor car – you must safeguard the health that
you have'.

Your duty, to yourself and those around you, is sensible
and quite clear. You must observe the regulations, carefully
follow safety instructions and use the protective equipment
that is provided.

Know hazards and prevent them

Take an extreme example: a lion walking down a street
unattended would be instantly recognised by most people
as a potential hazard and they would very quickly do something about it! In everyday work the sort of thing more likely
to cause injury would be an air line hanging across a gang-
way. You would instantly recognise it as a trip waiting to
happen. An electrical cable laying across the sharp edges of
a cut panel would, hopefully, result in immediate action to
remedy any possible hazard. Very regrettably, every year in
bodyshops, hundreds of people are injured, some maimed
for life and some are killed because the victim or their workmates did not take sufficient care. None of these people
wanted to be hurt, or to cause hurt. So how can you avoid
such things happening to you?

Try an experiment. For one week, or even one day, ask
yourself every time that you do anything at all, however
simple or trivial, 'could this hurt me or others?' If you do it
conscientiously, the number of potential hazards you find
will surprise you.

One major cause of injury and sometimes death is horseplay. Fooling about with equipment and pressurised gases is
extremely dangerous. One particular danger is that of compressed air. A number of people have been seriously injured
or have died as a result of compressed air entering the body
through an opening such as the anus. If a stream of compressed air or gas, even at a distance, is in-line with a body
opening, the gas will force entry and can rupture the passages inside, with all too easily disastrous results.

What you must report

Health and Safety regulations often make you just as
responsible as your manager or employer for compliance
with their requirements. That is especially important when it
concerns activity on the workshop floor. You and your colleagues know precisely what is happening minute by minute, so it is vitally important that anything of concern is
reported to your management immediately. Typical items or
events that must be reported include:

- Failure of any protective system, such as a dust extrac-
  tion unit.
- Defects in, or lack of, personal protective equipment.
- Faults with any service that could be a hazard, such as
  damaged mains electrical cables or equipment.
- Obvious hazards such as oil leaking from vehicles or
  equipment.
- Uncontrolled or uncollected waste.
- Any accident, however trivial.

- All injuries, including cuts and abrasions.
- Visitors without authorisation.

Many of these requirements also relate to any work carried out on the company's behalf outside the premises. In these temporary situations you should, of course, be taking proper care, but any event of whatever nature that may have later repercussions must be reported.

Health and Safety at Work in the Workplace

Emergency actions and first aid

One of your first actions when joining a firm or going to a new workshop is to get to know what first aid facilities are available and who are the first aiders. Find out the procedure in the event of an accident or fire and where the fire fighting equipment is located.

In the event of an accident, your safety and the safety of all those around you comes first. Only when there is no further risk to individuals should attention be given to the workshop and to vehicles.

Fire fighting equipment and its use

There are several elements to handling an outbreak of fire. Which equipment is used first must depend on the circumstances. However, always remember that lives are more precious than property. One of the most important pieces of equipment then must be the fire alarm, to warn all your workmates. Familiarise yourself with the procedure for alerting the company at large and those who will call the Fire Service.

You should know where the fire extinguishers are located, and which type is safe to use on the fire: an incorrect choice may spread the blaze and even cause you injury. Here are the types of extinguisher commonly in use and their correct colour coding:

- *Red*: water – for solids such as wood, paper and rags. Do not use on oil or electrical equipment.
- *Blue*: dry powder – all types can be used on liquids; multi-purpose is also suitable for solids. Safe on electrical equipment but as it does not easily penetrate the fire may re-ignite.
- *Cream*: foam – only aqueous film forming foam (AFFF) should be used and it is suitable for liquids and solids.
- *Black*: carbon dioxide (CO_2) – use on liquids and best for electrical.
- *Fire blankets* are good for solids, liquids and clothing, provided that the blanket completely covers the fire.

Importantly too, as with first aid, as early as practicable examine the fire fighting equipment and read the instructions for use. You can also get a copy of an extinguisher guide from your local Fire Brigade – there will not be time when there is a fire!

Removing or limiting risk

First, the risk of fire. Motor vehicles and the techniques and materials used in repairing their bodywork present a risk of fire. When a Fire Officer inspects your workplace, he will expect to see an operation that minimises risk and is complying with the requirements that are laid down. The main concern is that you, or anyone else, may be at risk of injury or death through fire. Some of the major considerations are as follows:

- Most Local Authorities have strict bylaws governing the storage of flammable materials such as paints. Only materials actually in use should be in the workplace and then only in the appropriate part.
- The greatest care must be taken with fuel systems. Carelessly fastened or sealed fuel pipes can very easily leak fuel. Tanks must never be emptied or removed within the main building: the Fire Officer will expect draining to take place where it can be done safely. Similarly, there should be a suitable store for fuel tanks removed from vehicles.
- Waste materials on the floor, welding equipment carelessly laid aside, inadequately protected flammable components such as trim, and general disorder will all pose risks to you and your colleagues.
- The fire alarm, evacuation procedure and fire fighting equipment must be to a satisfactory standard.
- To be able to evacuate the building, the Fire Exits must all be free to open, have clear access and an open space outside them.

There are numerous other risks of great variety. Tripping over cables, pipes or components may not in itself be very dangerous but with so many hard or sharp objects to fall against, the slightest loss of balance can result in serious injury. For example, keep cables and pipes that cross walkways to a minimum and on the ground.

Work itself, if carelessly carried out, may all too easily injure the operative or those nearby. Falling components, or even whole cars unbalanced by unit removal, flying particles from grinding, welding flashes and the body-pull attachment that tears away, can all be dangerous. These and similar problems can cause injury and even death.

So, what can you do about it? There will always be some risks but many can be removed or made much safer by the way that you work. Always ask yourself before beginning any task; what is the best and safest way to prevent any problems?

Here are some simple rules to help keep you safe:

- Have your own safety kit of personal protective equipment (PPE).
- Keep it in a container or locker when it is not in use.
- Keep re-useable items of PPE clean by wiping them over after use.
- Renew worn or damaged items of PPE as soon as necessary.
- The law insists that your employer provides PPE, and you must use it.
- Check your handtools – replace, repair or sharpen sooner rather than later. Loose hammer heads and 'mushroom' chisels are particularly dangerous.

- Major equipment deficiencies must be put right before use.
- Work tidily – only have ready on the job what is needed for the task in hand.
- If you see a hazard, deal with it or report it.
- If there is a safety procedure, follow it – it protects you.
- Think of those around you – warn and protect them as appropriate.

Who is responsible and communication

Everyone in the workplace is responsible and each must communicate with one another. There are though some specific duties for each member of the workforce. Here are some of them:

- RIDDOR (the Reporting of Injuries, Diseases and Dangerous Occurrences Regulations 1985) requires that all accidents resulting in three days or more off work are to be reported to the authorities. In addition, you should report all but the most minor accidents, *any* exposure to pressurised material or gas and any other abnormal event (see Fig. 5). Even if there is no immediate sign of hurt, the person who keeps the accident records in your company must be informed and an entry made in the Accident Record. Without such an entry, any future claim will be unlikely to succeed.

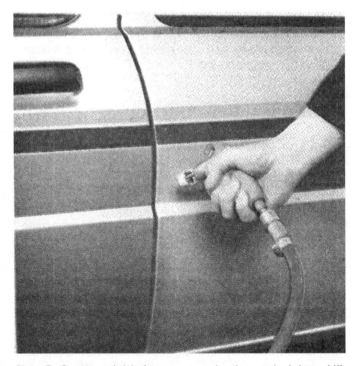

Figure 5 Compressed air is **dangerous** – pointed at your body it **can kill**; always use a controllable jet at the lowest pressure that will do the job, and blow across surfaces, never directly at them

- Incidents involving pressurised materials must be taken very seriously, even if there is no immediately visible damage. The loss of a limb or death can occur if any fluid, semi-fluid, gas or air enters the body. This may happen in

two ways. First, by accident or horseplay, which can occur when overalls are blown off with an air line. As overalls offer no protection against pressurised gases, a jet of air can enter the anus. As the colon can rupture at a pressure as low as 6psi, serious injury can easily occur and may be fatal.

Secondly, horseplay again or normal work may also inadvertently expose the outer skin to a high-pressure jet at a short distance. There may be few if any marks on the skin but internal damage can lead to gangrene and the loss of a limb. This is caused by gas or fluid passing through the tissues and even, sometimes, entering the blood stream.

Any incident involving pressurised materials or gases must be referred to an Emergency Accident Unit at once and be reported in your company, and to the authorities.

- Keeping those around you informed of what you are about to do is also very important. A small group working together becomes used to the activity going on around them and, almost by instinct, will look away from welding flashes, keep clear of a body pull or cover their ears against noise. But do not rely on this happening. It is not always convenient when small amounts of such work are involved to move screens into place. In such circumstances make sure that everybody is told what is about to happen. If personnel from other parts of the company are likely to enter, warning signs should be displayed and screens moved into position.

Cleanliness is Health and Safety at Work

Benefits of a clean workplace

A dirty and untidy workplace displays to everyone that nobody cares – about the company, the customers or, perhaps more importantly, about themselves. It will almost certainly be a place full of hazards, where dust and fumes are part of the air that is breathed (see Fig. 6), noise affects hearing, where the bodies of the workers slowly deteriorate and activity causes injury much too frequently.

A clean and tidy workplace is one where care for everything and everyone is paramount. There is a very good chance too that it will be more profitable, both financially and certainly physically. Your working life should be rewarding and you should have pride in your work and in your workplace. Leisure time can be enjoyed and you will hopefully be fit for retirement when that day eventually comes. A pleasant workplace demands commitment and co-operation from the company and, of course, from each and every member of the workforce.

Do remember that your environment is, to a large extent, what you and your colleagues make it.

Cleaning requirements

Cleanliness of the workplace brings very obvious and positive benefits for both the paintshop and the panelshop. The

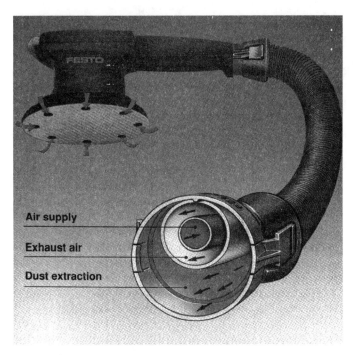

Figure 6 Dust extraction and one connecting pipe to minimise hazards

paintshop benefits through low dust levels, enabling a high-quality paint finish to be achieved with less effort. For the panelshop, the reduction of hazards brings about safer, faster and easier work.

Cleaning amounts to much more than, say, sweeping the floor and wearing clean overalls. Regrettably, few body-shops take the trouble to clean vehicles before repairs commence. To do so would minimise the accumulation of road dirt and oily deposits and help make for clean, fast and safe work. There will, however, always be a considerable amount of road dirt and dismantling debris when bodywork repairs are carried out. The secret is to clean up frequently. It takes no longer to 'clean up as you go': the dirt will not be trodden down, the fire risk is reduced and you will be working in cleaner conditions.

Anti-corrosion warranties also demand a clean approach. Steel particles from grinding or welding, if left in body recesses and box sections, are likely to corrode and may well cause failure of the section in time. With warranties of up to ten years, there is plenty of time for corrosion to occur. In such circumstances, the repairer is liable, not the car manufacturer. Prevention is always better than cure so always vacuum-out any debris.

Preventing soiling and damage is vital in the case of interior trim and components. Adequate covers should be used inside the vehicle, of fireproof type where appropriate, to eliminate or at least reduce the cleaning or replacement that may be necessary after repair.

The one topic that has not so far been mentioned is a most important one – you! Many will laugh at the thought of being clean when working in a bodyshop. But that should be your aim. The processes and materials in use for much of the time are alien to the human body and it must be protected from them to maintain health and long life. We have already considered physical damage to the body, including dangers that affect hearing and sight. What of the less obvious dangers to your body's internal systems?

Danger can very easily arise from what is drawn in when breathing, or ingested when eating or drinking. Did you know that it is an offence to eat or drink in your workplace? It is also an offence, to your body and its health, to eat without previously washing your hands (see Fig. 7). The effects of minute doses of poisons over many years can eventually be dramatic. Yes, it is a nuisance to have to wash your hands frequently, but personal cleanliness is essential for good health. Gloves of all types are available and should be used all the time. It is possible to work in gloves – the eye surgeon and the dentist do the most detailed work in gloves, so I am sure that you can – if you try.

Equipment and consumables

For routine gathering of dirt and dust, and picking it up, the old fashioned broom and shovel take a lot of beating. Industrial vacuum cleaners are more cumbersome but very efficient, provided that they are emptied from time to time! They can even be too efficient. If you forget to put away some screws or clips then they too might be picked up. Vacuum cleaners are excellent for cleaning out box sections and removing debris from the job. Some dust extractors, too, can be used in the cleaner mode. Absorbent granules or a similar material are useful for soaking up liquid spillages.

General cleaning equipment can also be used to remove loose dirt from body jigs, hoists, installations and paint booths. Sealers or other sticky residues should be removed with a suitable solvent. The appropriate safety precautions must be observed and personal protective equipment used as necessary. Any machinery must be isolated from the mains or rendered inoperative before any cleaning. Disposal of all waste must follow the correct code of practice.

Regulations

Cleaning and the end product, the rubbish or waste, is covered by more than one set of regulations. The COSHH regulations will apply to much of the act of cleaning away dirt and waste. The protection used in creating the waste in the first place should, broadly speaking, be employed in handling it and in using solvents and cleaning agents of a toxic nature. For example, a painter who correctly used solvent proof gloves to mix paint would use them too for disposing of the used wipes (see Fig. 8).

Handling and storing waste will also involve compliance with the Control of Pollution (Special Waste) Regulations 1980 and observing the Duty of Care under the Waste Management Code of Practice 1991. In practical terms this means storing the waste in separate, adequate containers, secure against wind and weather. Waste from the paintshop in particular should not be allowed to accumulate inside the main building as many paint products can, in certain circumstances, self-ignite.

These regulations and other requirements are the reason why you are asked to separate your waste into categories and not to mix them. Liquid waste from the paintshop is handled quite separately from 'dry waste' which may also be split into various categories if recycling takes place. One final point: do remember that the regulations provide for fines on personnel, as well as the company, where a breach of the code of practice has occurred, even though a waste disposal offence has not been committed!

Dirt on the skin

1 Cleansing agent covers the dirt particles = emulsification

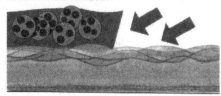

The dirt is washed off with water

2

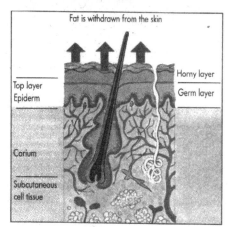

3

4

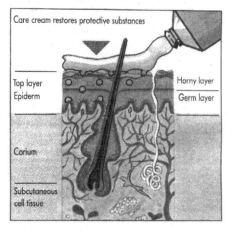

Figure 7 Caring for your hands:
1 dirt particles on the skin emulsified by a cleanser and
2 washed away with water
3 fat is lost from the skin so
4 it is replaced with an industrial hand care product

Figure 8 Look after yourself in the mixing room too!

Chapter 2

Finding Information

In times past it was possible to carry out most repairs to the common marques of motor vehicles with little or no need to consult service information. That is certainly not true now. When you work on a present day vehicle you will need information for most, if not all, of these reasons:

- the variations in vehicle design,
- the need for accuracy in alignment and adjustment,
- a wide variety of materials,
- different methods of assembly,
- the requirements needed to meet safety features,
- anti-corrosion warranties and
- legal obligations.

All these reasons mean that there is a heavy responsibility on the shoulders of those reinstating accident damaged vehicles. To this must be added:

- the complexity of the equipment and
- special materials used in repair and refinishing.

Now, accurate information is a necessity and those who attempt repairs without it may so easily find themselves in difficulty or even facing litigation.

Your work involves reinstating and preserving from corrosion the life-saving features built into a modern motor vehicle. The quality of that work may make the difference between life and death for the driver and passengers in the event of another accident. Only by knowing and using the correct materials, equipment and procedures can you be sure of a good job.

Legislation

Copies of regulations and guidance notes are available from HM Stationery Office, Government book shops and the Library and Information Service of the Health and Safety Executive. Your college library may well have copies that you can study. Retail book shops are usually quite happy to order publications but you will need a reference number. An enquiry number for the Health and Safety Executive can be found in every telephone directory.

Trade magazines are also very helpful. The monthly periodical, *Bodyshop Magazine*, frequently carries detailed articles on body repair topics. It also has supplements on specific subjects that contain a wealth of information provided by companies specialising in that field. The guidance given on legislation, and the equipment and consumables that are necessary for its implementation, are particularly useful. *Body*, the journal of the Vehicle Builders and Repairers Association (the VBRA), may also carry some technical information from time to time.

Vehicle manufacturer's information

All vehicle manufacturers produce service manuals for their products but manuals on body repairs and refinishing vary greatly from franchise to franchise. Normally, if body repair manuals exist, they are restricted to the manufacturer's dealer network or to approved specialist body repairers (see Fig. 9). Most manufacturers provide a means of bringing these publications up to date. This is usually in the form of bulletins or service letters.

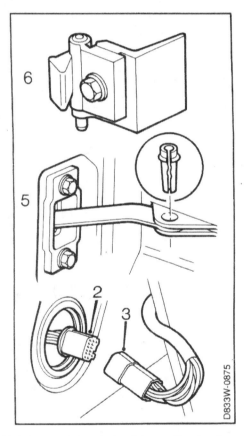

Figure 9 Workshop manuals and microfiche contain many illustrations, such as this example for the Saab 900; the numbers refer to instructions in the manual

Another very useful source of information is the parts list or parts microfiche. These usually have exploded illustrations and can be most useful if you need to know the separate components in a group and their order of assembly (see Fig. 10). If you do not have access to them and you need such information, the partsperson at the local dealer would probably help to resolve a problem.

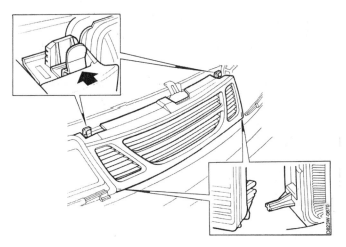

Figure 10 Some illustrations are so clear that text is hardly needed, as in this example from the Saab 900 Body manual

Training courses and distance learning materials such as video programmes are also often made available or organised by the manufacturers or importers. These may be identical to those offered to the trade at large, but sometimes are partially or wholly specialised (see Fig. 11).

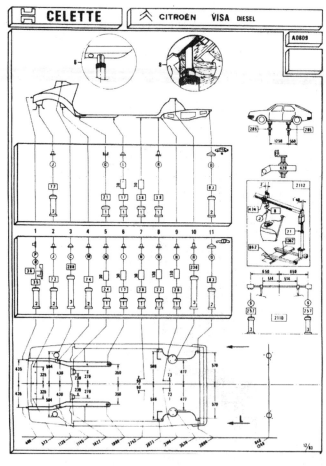

Figure 11 Body measuring systems demand clear setting-up instructions and precise dimensions

Vehicle manufacturers also often supply equipment and consumables or specify which particular makes to use.

Back-up services provide guidance on equipment and consumables that are generally available and suitable for their vehicles.

Where refinishing products are concerned, most makers specify one or more brands of paint. These are often made by the companies who supply materials in bulk for the production lines in the factories. As a result, information on correct refinishing techniques is available to all users from paint suppliers.

MIRRC information service

The Motor Insurance Repair Research Centre, more familiarly known as 'Thatcham' after the name of the town in which it is based, provides perhaps the most comprehensive information service to its bodyshop subscribers. It was set up by the British insurance industry with the aim of minimising the cost of accident damage repairs to motor vehicles. It is to the credit of those who run the centre that their activity is aimed at helping the body repair industry to produce the highest quality of repair at minimal cost. In this way the integrity of the body structure is retained and the manufacturers' warranties can remain valid.

Thatcham's activities have extended to meet the needs of both the insurance companies and the body repair industry during a period when body design and construction has changed. During the last twenty years or so safety features such as crumple zones have become commonplace. At the same time vehicle makers have been trying to make their products lighter and so more economical on fuel. Anti-corrosion design and treatment have made modern car bodies much more durable. All these factors have influenced the nature of accident damage, the quality of the repairs and the methods employed to carry them out.

These facts and the pressure to keep down costs have created a need for training and other forms of information that will help the bodyshop and its staff to remain competitive. Thatcham provides services in the form of courses on estimating and repair techniques, instructional video programmes and the invaluable repair manuals. Some of the information contained in them is provided by the manufacturers, but most is based on practical experience gained from doing the job themselves and this also has the full approval of the vehicle manufacturers (see Fig. 12). They also have the widest collection of currently available equipment in one place and their staff use it everyday. It is also noteworthy that the bodyshop staff at Thatcham always use every means of personal and environmental protection available to them when working. Their workplace is a good example of health and safety being practised at work.

Equipment manufacturers

Few bodyshops have all the equipment that their staff would like. It is also true that few bodyshops make the fullest use of the equipment that they do have. This may be due to poor condition because nobody maintains it, or because those who use it are not fully aware of what it can do. The information provided with equipment is invaluable in getting the best from it. The data supplied with body jigs for example, is

INTRODUCTION

VEHICLE SAFETY FEATURES

Precautions

When carrying out repairs to these vehicles where it is necessary to disturb the airbags or seatbelt grabbers in any way, it is essential that the following procedures and precautions are observed:

- Disconnect **both** battery terminals and insulate them thoroughly. Wait for at least **10 mins**. before proceeding further.
- Disconnect the multiplug connector at the control unit (*see Fig. 13 page 16 and Fig. 9 page 12*).
- Remove the airbag module retaining bolts, disconnect the airbag inflation device, and place the module in a secure storage area with the cover facing upwards. The storage facilities must comply with the appropriate Regulations (*see Thatcham Report 'Airbags' for further information*).
- Following an accident it is essential to renew all SRS components. Do not attempt to repair faulty or damaged components. Where the airbag module is renewed, the new expiry date sticker must be fixed in place over the original sticker. All airbag modules must be repaired automatically after 10 years service, and all identification numbers on both new and old modules recorded from new in a register along with the other vehicle details.
- SRS components must be tested for correct operation using the Manufacturer's diagnostic equipment, following replacement. Where an airbag has been disconnected with the ignition switched on this will register in the electronic memory as a fault, which will necessitate resetting the system after reconnection.
- Health & Safety Regulations must be observed when handling an airbag module, whether it has been activated or not. When handling a module which has been triggered wear protective gloves and wash exposed skin with a neutral soap. Airbag modules **must be renewed** if they have been accidently dropped from a height greater than **750mm**.
- Where a vehicle with an unexploded airbag has to be scrapped, the module must be triggered electrically as a safety measure prior to disposal. **Do not attempt to scrap the vehicle before performing this operation.**

- Do not attempt to work on an activated airbag for several minutes following activation, as the metal components will be very hot to the touch and could cause personal injury.
- Do not expose an airbag module to temperatures reaching or exceeding **85°C**. Never allow naked flames near an airbag or any other part of the Safety Restraint System.
- When working near seatbelt grabber assemblies do not use high impact loads (e.g. from a hammer) within a radius of **600–700mm** of each assembly. Where high temperatures are to be applied during paint curing operations it is essential to remove the arming rod from the assembly at **2** in **Fig. A** after removing the screw **1**. **The arming rod must only be subsequently refitted with the seatbelt assembly in place on the vehicle.** Do not expose a seatbelt grabber unit to temperatures greater than **110°C**, and always remove the arming rod before dismantling the unit. **In cases where a unit has been activated it must be renewed.**

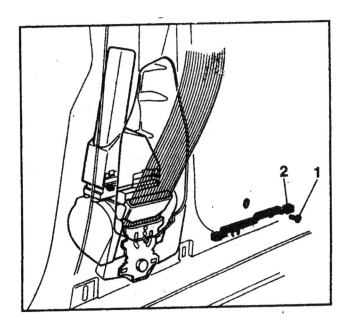

Figure 12 MIRRC Methods manual information: vehicle safety features

very comprehensive and because it is an essential aid in using the jig, it is usually available in the bodyshop.

That is often not so where other items of equipment such as welders are concerned. All too often, the manual is laid aside, without even a glance at its pages, because everyone thinks that they know how to use it. That is a normal human reaction: we like to discover for ourselves how to use things. Unfortunately, important facts often remain undiscovered, sometimes for ever, unless someone reads the manual.

Many makers of equipment provide formal training courses or familiarisation sessions on-site. They are also quite willing to answer questions or to give guidance or advice if there is a problem that needs solving. Always ask if you have a query: the answer is often quite simple.

Material manufacturers

The makers of materials used in body repair are amongst the most helpful companies that there are. From sealers to paint systems and from masking to PPE there are very few who cannot offer an information service of the highest quality.

The paint companies offer the most sophisticated information, as the nature of their product demands it. With few exceptions, vehicle manufacturers are happy to leave the supply of information and on-site guidance in the hands of their chosen suppliers. If you are a painter you have one of the best training and information schemes at your disposal and for most paint makers it is true to say that the answer to any problem is just a phone call away. You will also know that a good part of your success will be down to how well you can use the information provided.

Some of the other makers of consumables are not far behind. Those providing masking materials, polishes and abrasives, for example, are also good with both advice, on-site guidance and even formal training. Information, usually in leaflet form, is available from most of them. Once in the bodyshop, leaflets are usually lost after a short while: why not ask for them to be filed in a special folder for future reference?

Suppliers

Major suppliers of consumables, the factors, also provide useful information on the materials that they sell. Much of this can be used to check if you are using the best or most cost effective materials for any given job. It will also keep you up-to-date with new products or services.

Computerised estimating systems

These systems are primarily designed to speed up and make more accurate the time consuming job of estimating. One facility that may be available is that of repair information, including cutting lines for panel removal and exploded views of body components. Although the information is usually sourced from vehicle makers, the possibility of recent modifications should always be borne in mind.

Chapter 3

Replacing Body Panels and Fittings

Dismantling for Repair

Health and safety at work and personal protective equipment (PPE)

The checklist below reminds you of the hazards of day-to-day work and those that can arise when dealing with major jobs such as bodyshell changes:

Hazard	Protection
Damaged metal and plastic components often having sharp and jagged edges Shattered glass resulting in sharp slivers	Riggers' gloves Goggles or face visor to BS 2092 Grade 2 or EN 166 grade 'F' Industrial overalls
Liquids such as coolant, oils, hydraulic fluids and battery acid; the main danger is involuntary ingestion (taking in through the mouth) – this may occur through splash but is more likely through handling food with unwashed hands	Wear suitable gloves A face shield should be used if there is danger of splash from pressurised fluids Always wash your hands before attending to personal needs Food and drink must be consumed in a designated area
Dust may be generated in small quantities when revealing welds	Dust mask to BS 6016 or EN 149 class FFP1
Brake fluid is very poisonous – just a teaspoonful is fatal Always check the contents of an opened drink can before drinking; someone may have bled brake fluid into it	Never take drinks or food into the working area Throw used containers away, preferably crushed
Some airbag systems, usually American specification, are fitted with a delay device which allows activation for up to 30 minutes after the ignition has been switched off	Keep out of the activation path of the airbags If in doubt wait 30 minutes after battery disconnection
The active component of airbags, sodium azide, is very dangerous Unburnt propellant which has escaped from a damaged module can ignite at 200 degrees Celsius, by friction or pressure and can produce toxic gas if mixed with acid	In the unlikely event of spilt, unburnt propellant, it must be gathered into a sealed container by a person wearing full breathing apparatus No other persons should be allowed into the area Never cut or heat an airbag inflation module
Bonded glass may sometimes be removed with a wire cutting tool These wires are capable of inflicting injury if they break	Heavy leather gauntlets and eye protection should be used, together with overalls
Flying particles, particularly metal, from grinding and cutting	Goggles or visor to BS 2092 Grade 2 or EN 166 grade 'F' Industrial overalls Riggers' gloves
Pressurised gases such as compressed air can be extremely dangerous if used in an uncontrolled manner	Compressed air and gases must only be used with equipment that provides full shut-off and directional control Normal workwear provides no protection against pressurised gases
Gas R12 from air conditioning systems is dangerous because it becomes toxic when burnt, can cause cold burns and, as it is heavier than air, will flow to the lowest point	Gas should be evacuated into a reclaim unit for processing Wear heavy, heat insulating gloves Work should not be done in a workshop with a pit – Escaping gas is heavier than air, and can asphyxiate anyone there

Hazard	Protection
Note: Gas inhaled through a burning flame, such as a cigarette, is converted to mustard gas, a deadly poison.	
Mains electricity is as dangerous as pressurised gases if used carelessly and the equipment poorly maintained	Plugs, cables and equipment must be in good order without defects Cables and plugs must be protected from cuts, chaffing, crushing or any form of tension
Injuries to feet from heavy components, sharp metal or hot particles	Footwear to BS 1870 (200 joules), preferably with steel midsole
Noise from cutting, drilling and grinding equipment	Ear defenders to BS 6344 Part 1 Warnings to others in the vicinity or noise screening
Crushing injuries from the vehicle or equipment	Never place any part of your body between elements that could move together Always ensure that vehicles are fully supported and cannot overbalance

Vehicle protection and housekeeping

The customer's car should be protected from the dirt and dust that is generated during body repair and from the risk of damage or fire from hot metal particles and welding. A clean working environment will largely take care of the dirt and dust element but it is good practice to protect trim and seats with suitable covers anyway. The front seats should have been fitted with seat protectors on arrival, of course, unless they are to be replaced.

During repairs any remaining interior trim should be protected in a manner appropriate to the work. During normal activity conventional dust covers are quite adequate. When hot metal particles are being generated and during welding these should be supplemented or replaced with fireproof blankets.

The outside of the vehicle must also be protected, particularly from the minute metal particles that fly about during cutting, drilling, grinding or sanding. These can collect in crevices and subsequently cause corrosion. Hot particles from these activities or from welding can sink into the paintwork and they, too, will corrode in time. Those who work in premises beside railway lines near to stations will be familiar with the way hot particles from the train brakes settle on cars parked in the open. These impregnations can be removed with dilute oxalic acid but the work is slow and tedious. It is much better to prevent the contamination in the first place.

One part of the car that is often overlooked is the wheels. Tyres are a very important part of the safety equipment on any vehicle and should be treated accordingly. The best way of protecting wheels and tyres is to remove and store them. This is not always possible so they should always be covered by stout covers. If there is likely to be waste on the floor that could damage the rubber, the wheels should be lifted clear while the work is in progress.

Many of the points on vehicle protection are part of good housekeeping: the activity that removes waste and rubbish now – rather than later. Controlling the waste you make and cleaning it up immediately gives you a cleaner and more comfortable working environment. Above all else it gives you a safer place in which to work. The chances of tripping or slipping are reduced and other potential hazards can be seen easily.

Keeping the vehicle and the workplace clean helps with personal hygiene too. You are much less likely to be harmed by toxic substances, either by skin contact or by ingestion (that is taking them in through the mouth).

Sources of information

It is worth spending some time looking for information – it often saves hours of unnecessary effort. Here are some suggestions for each type of job dealt with in this chapter:

Job	Source of information
Removing mechanically held components (Internal and external trim, electrical components, mechanical units and bolted panels)	Vehicle makers' information (manuals, parts lists, training material) MIRRC (Thatcham) information Proprietary manuals
Removing bonded glass	Vehicle makers' information MIRRC information Equipment makers' manuals Material suppliers' data

Job	Source of information
Removing bonded trim	Vehicle makers' information MIRRC information Proprietary manuals
Removing welded panels and components	Vehicle makers' information MIRRC information Equipment makers' literature Training information
Cleaning and waste control	Equipment makers' information Material makers' or suppliers' information Company safety policy

Trade journals such as *Bodyshop Magazine* often carry articles or supplements on elements of the work in a bodyshop. These can be invaluable, especially in keeping you up-to-date with tools, equipment, materials and techniques.

Equipment

Hand tools are all that is needed for the work in this section with the possible exception of lifting equipment. Work is always easier, quicker and, because it can be more easily controlled, safer when a vehicle can be lifted to a convenient working height. It is also safer because it is much less tiring. If it is possible to raise the vehicle to a better working height safely, then do so. The vehicle must be safely supported and stable, of course. Never rely on a trolley jack without axle stands.

Basic hand tools are a variety of screwdrivers or a cordless powerdriver with a range of bits, spanners and sockets. A smalldrive socket set with a hinged ratchet will be very useful. Trim removal can be helped a great deal by the specially thin removal tools made by several major tool firms. A good bolster and a small cold chisel are needed for panel separation.

The removal of bonded glass will call for the use of a special tool (see Fig. 13). There are a number available ranging from wire cutting, sometimes using winches, to various types of oscillating cutters (see Fig. 14). Suction pad glass holders are very useful for lifting the glass.

Removing body panels calls for the use of various cutting tools. Drilling out spot welds is easiest, with less chance of drilling away the base panel, using an adjustable drilling attachment. Special drill bits are also made for this work. Panels are best cut with a pneumatic oscillating saw. An angle grinder is useful to cut through folded edges.

Storage of components can be catered for by making or purchasing a trolley with compartments and containers.

Methods of construction

The thin sheets of steel from which most car bodies are made are very flexible and have little rigidity. You can test the principle by which this apparently weak material can be made into strong car bodies by carrying out a simple experiment.

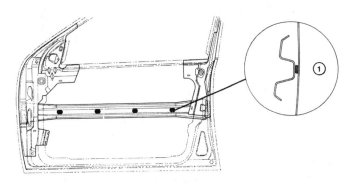

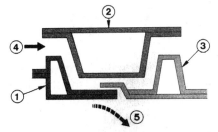

Source: Reproduced by kind permission of the Ford Motor Company Ltd.
Figure 13 (a) Door reinforcement on the Ford Mondeo and the bonding points
(b) The overlap of the front door (1) on the Ford Mondeo ensures that it cannot jam as a result of extensive body deformation in the direction (4); the door will move out from the B post (2) and the rear door (3) in the direction (5)

Make a tube by rolling up a sheet of ordinary note paper. Then glue the overlapping ends together. Stand the tube on end on a level surface and put a disc, like a coaster, on the top.

A considerable weight can be stood on top before the tube will buckle. As with sheet steel, the paper is decidedly floppy when unshaped but remarkably strong when it is formed into a three dimensional object. You can imagine, then, how much strength steel sheet will have when formed into such shapes.

The object of the car designer is to obtain the strength needed from the least weight of metal. As you are already aware, the individually shaped panels and other parts of a vehicle body are joined together in a variety of ways. Folded edges, various forms of welding, bonding, clinching, riveting and bolts and nuts are all used. We will be looking at how you recreate most of these joins to repair the body later in the book.

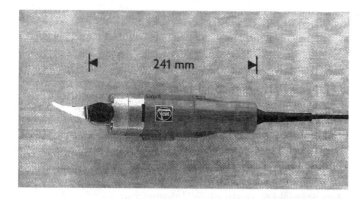

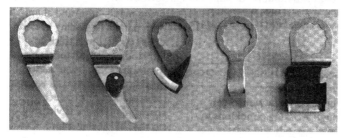

Figure 14 (a) A popular, compact oscillating blade cutter for bonded glass removal
(b) A selection of the cutting blades available; the chisel cutter on the far right is ideal for cleaning off excess sealer

To understand present day body construction it is useful to consider how the methods used have come about. In the earliest days of motoring, cars were little more than converted horse-drawn carriages with an engine instead of horses to move them along the road. When cars were first designed the accepted method of construction was still followed. A ladderlike frame was made, called the chassis, onto which were assembled the power unit and running gear. A completed body was then mounted onto the chassis. This method was followed for many years until people such as Henry Ford started mass production. This led to an integrated form of construction called monocoque, using pressings fastened together. Gradually, all high-volume production was changed to this method although many commercial vehicles are still made with separate chassis, cab and body. Only a few specialised cars are made with a chassis today.

One other variation on the chassis–body theme is the use of a platform frame. Here the floor of the body forms the platform to which all the other components are attached. The classic vehicle made in this way is the Volkswagen Beetle.

Mass produced cars and small vans have been made in the monocoque design for many years. In this type of construction the body sills, underfloor members, extensions at the front and a strengthened luggage boot replace the separate chassis. The other body panels, and often the glass too, all contribute to the overall strength and stiffness of the body. This method is ideal for mass production.

Another way of making car bodies is to use a space frame. In this construction a frame is made in the shape of the vehicle. Outer panels are attached to the frame to complete the body. In early designs the panels were purely cosmetic and provided no strength. The space frame is in use again in some of the latest designs of aluminium bodywork. But in these the panels and the glass now provide some of the strength.

For many years, body parts have been joined together on the production lines by welding, mostly by resistance spot welding. Thicker materials or joints carrying greater strain may be MIG/MAG welded. Some makers have also used a little brazing. More recently, bonding with adhesives has been used for some joints, usually combined with a folded flange or spot welding. Glazing is often bonded, mainly the front and rear screens. Outside trim may also be attached to the body by adhesive or adhesive tape. Bolts, screws, rivets and clips continue to be used for attaching all the other parts to the body shell.

Methods of jointing never previously used on cars are also employed on the new generation of aluminium bodies, both on the factory production line and in the bodyshop. Because the spot welding of aluminium needs much more power, one alternative method now being used is punch riveting. It is claimed to be stronger and uses less power. Clinching, a technique where the parts are pressed into a locking shape every so often, is also being used (see Fig. 15).

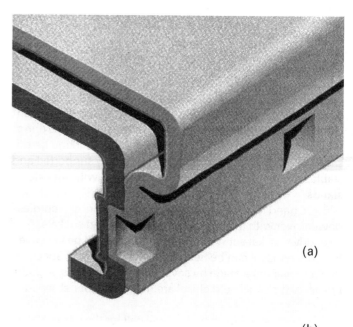

(a)

(b)

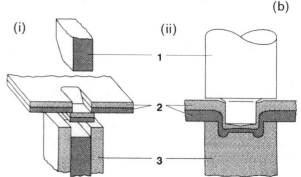

Source: The drawings at (b) are reproduced by kind permission of Robert Bosch GMBH.

Figure 15 (a) A section of a clinched joint on the all-aluminium body of the Audi A8; not as strong as spot welding, this method is only used for panels of secondary importance
(b) This diagram illustrates clinching (i) and Tox clinching (ii) the advantages are the absence of heat and the ability to join totally dissimilar materials such as steel and plastic (2). (1) and (3) are the press tools

For several decades, safety conscious manufacturers have sought to improve the effectiveness of the monocoque body and its 'passenger safety cell'. In the early days they concentrated on the absorption of front and rear impact energy and the effects of the 'roll-over'. Latterly, side impacts have become the subject of much discussion. All these safety features must be kept in mind when you repair or modify a car.

In any impact, protection of the occupants is best achieved by limiting sudden stresses on the human body and by protecting it from violent contact with solid structures. The protection from frontal impact is achieved by building into the car 'crumple zones' or impact absorbers, the car version of railway buffers (see Fig. 16). These are incorporated at the front and rear of the car bodywork. As the car strikes a large, hard object or another vehicle, the front or rear of the car will collapse with gradually increasing resistance until, in the most severe impacts, it is pushed back to the passenger cell. This gives a slowing down effect that can prevent severe injury to properly restrained occupants. At the rear, the spare wheel is sometimes laid flat in a well to provide part of the compression resistance. You can see then that such spare wheels should always be properly inflated and never left out!

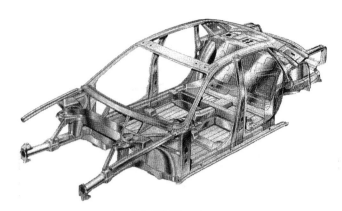

Figure 17   The space frame of the all-aluminium body on the Audi A8

Although the side of the vehicle will come into contact with the occupants, a good proportion of the inertia will have already been absorbed in the collapse. Whichever method is used, padding and even side mounted airbags are now being used to help soften any impact with the bodies of occupants (see Fig. 18).

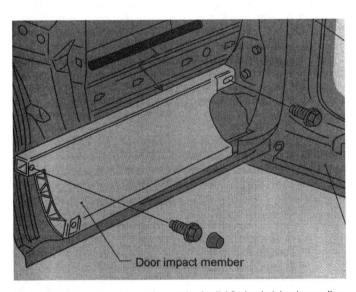

Figure 16   A door impact member on the Audi A8; the dark bar is an adhesive butyl cord which prevents noises between the member and the door

In the 'roll-over' situation, or when a heavy object falls onto the roof of the car, the structure is designed to withstand greatly increased pressure to protect the occupants (see Fig. 17).

Side impacts are more difficult because of the close proximity of the car body structure to those inside it. This reduces the amount of crumpling that can be used to overcome the inertia of the impact. At the time this book was written there were as many reasons for not having side impact protection as there were in its favour. Generally, side protection takes the form of a beam part way up the door and even stronger sill members, door pillars and roof rail. Such measures provide considerable rigidity. But because it is difficult to absorb the inertia created by the accident, the occupants may be thrown hard against the structure. Others feel that it is better if the side collapses inwards some way.

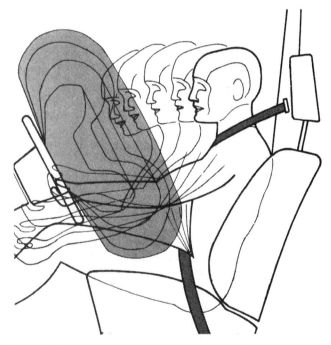

Figure 18   'Everything happens in a tenth of a second'
A typical sequence of events in a frontal impact
10 ms – airbag triggers
25 ms – driver starts to move forward
40 ms – airbag is fully inflated
60 ms – driver sinks into airbag (seatbelt stretched and tensioned), airbag starts to deflate
100 ms – vehicle speed is zero
110 ms – driver moves back into seat and airbag almost fully deflated to give clear view again
100 milliseconds (1/10 second) is the time it takes to blink

The safety measures used by many car makers, particularly those in Europe, rely for their effect on properly restrained and supported occupants. That means correctly adjusted seats and back rests, fastened and tightened seatbelts, and properly adjusted head restraints. Correctly used, the occupant

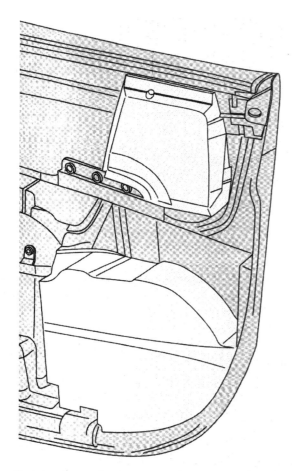

Figure 19 Protective padding in the doors on the Audi A8; the upper protects the ribs and the lower the pelvis

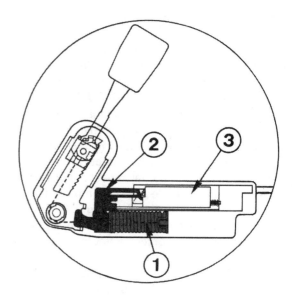

Source: Reproduced by kind permission of the Ford Motor Company Ltd.
Figure 20 The spring-activated seatbelt buckle tensioner on the Ford Mondeo
1 Cylindrical coil spring
2 Lever system
3 Spring mass sensor
The spring sensor releases the coil spring by moving the levers
The unit cannot be reset and must be replaced after activation

protection measures built into vehicles have been very successful in reducing injuries (see Fig. 19).

Car makers are trying to avoid dependence on the user by automating some of the actions and providing additional features. Seatbelt tensioners (see Fig. 20), including explosive types, are now quite common, along with devices such as airbags (see Fig. 21). To soften or avoid the injuries from seatbelts themselves, tear loops are also provided to allow some 'give' under high tension (see Fig. 22).

All of these features come under the heading of 'passive safety' – they minimise the consequences of an accident. When repairing the vehicle it is your responsibility to rebuild the structure as it was originally, consistent of course with the age of the car. You also have responsibility where 'active safety' is concerned – the steering, suspension and braking systems that help to prevent accidents. The correct alignment of steering and suspension is often in the hands of the bodyshop because there are few adjustable components. Failure to rebuild the vehicle correctly could result in occupants sustaining serious or even fatal injuries in any future accident.

Storage of reusable components

Reassembling a vehicle after structural repairs and refinishing often takes longer, and is made difficult, because components are damaged or missing. This causes delay and loss which can be expensive for both the bodyshop and the customer.

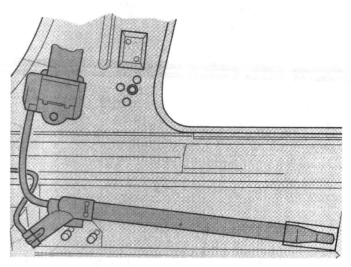

Figure 21 A pyrotechnic seatbelt tensioner; it is activated by the airbag control unit

To overcome such problems, components should be stored in wheeled storage racks (see Fig. 23). These should be padded for items easily damaged or scratched and be provided with containers for clips and other small items. Bag loose screws and clips and attach them to the unit or tape screws into position. Another method is to push them through a piece of cardboard, making a note of the component or location. If you discover a sequence for dismantling and reassembly, make a note of it and attach it to the part.

List damaged components as they are removed, especially clips and screws, so that replacements may be ordered and are available for reassembly. The important thing to remember is that a minute spent now will save ten at reassembly time.

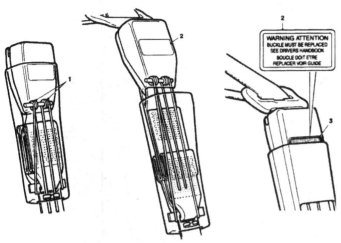

Figure 22 Some seatbelts used in conjunction with airbags are designed to 'give' a little at pre-determined loads to avoid injury; these Jaguar belts must be renewed as soon as any part of the label becomes visible; seatbelts worn during an impact must always be replaced

Figure 23 Ever wasted time looking for mislaid parts? Caring for the customer also means caring for the car and its components by using a component stand; this is a proprietory unit

Larger components such as windscreens can be stored in a special rack fixed in a safe place. Suitable racks can be purchased but are easily and inexpensively made from strip metal and board.

Removing internal trim

It seems impossible that any new way can be invented for holding interior trim into place. Yet very often, new models do have different types of clips or new fastenings for door and window controls. The first time that you strip out such a car can be interesting, to say the least, especially if items of trim and their fastenings are not yet available as parts (see Fig. 24).

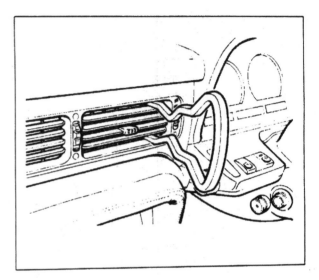

Figure 24 Sometimes special tools are needed to remove internal trim, such as this facia vent removal tool for the Jaguar

As with outside trim, information on the dismantling procedure is best. If its not available consider how the component could have been fitted. Screws will be out of normal sight, say on a hidden flange. If they can be easily seen, the heads will be covered by a cap in a recess. Hidden retainers are either lugs moulded into the trim and engaging with openings in the body members, or clips linking the trim to the body. For a first movement try sliding the trim from side to side or up and down. Such fastening usually moves relatively easily. If this is unsuccessful, the trim will almost certainly pull outwards.

Door and window controls may also be a source of frustration. A variety of screws, parallel pins and clips seem to cover most models (see Fig. 25). Check first for the fastenings you know and, if unsuccessful, try to work out how it may be attached. Door panels and other large items will usually pull outwards from the bottom with a slot or clip-over arrangement at the top.

Removing external trim

Removing items of external trim without damage on a car that you have never seen before is much the same problem as that posed by the internal trim (see Fig. 26). You need detailed information or plain old fashioned intuition!

Plastic components, such as bumper assemblies, are often clipped together and held to the body by clips and screws (see Fig. 27). Wrap-around units are best lifted away by two people to avoid scratched paintwork. Sidetrim is often held by adhesive or adhesive tape. A hot air blower will often help to release the adhesive. When using hot air, take care that hot spots are not created on either the components or the body; the trim may need replacing or other damage rectified at your company's expense. Other items are usually bolted through the body or into captive nuts. The search for cleaner body shapes giving less drag has resulted in covered gutters and sometimes 'A' pillars. Metal, plastic and rubber covers will be held by clips or studs.

Be careful of special components like variable aerofoils. These will probably be driven electrically. Sun roofs, too, must be dismantled carefully to prevent damage (see Fig. 28).

Bring to the attention of your supervisor incorrectly fitted accessories, especially in the case of tow bars. Your company will need to bring damage or defects to the attention of the owner or the insurance company to avoid being held responsible for rectification.

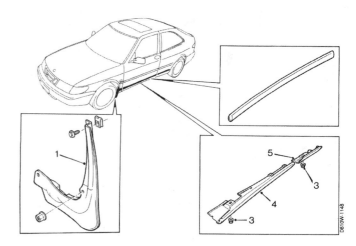

Figure 27 Splash guards and scuff plates secured with plastic nuts on the Saab 900; the numbers refer to the text in the manual

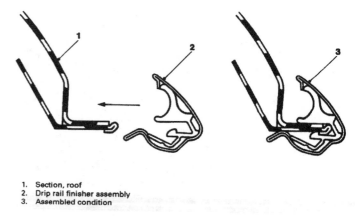

1. Section, roof
2. Drip rail finisher assembly
3. Assembled condition

Figure 28 Jaguar drip rail finisher

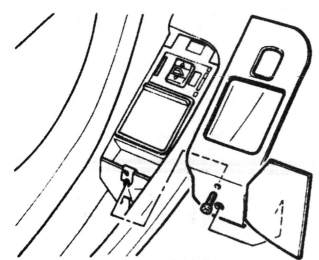

Figure 25 Door window switch trim on the Jaguar

Removing electrical components

Vehicles needing body repairs should have their airbags disconnected to prevent them inflating accidentally. This should be done before any other work is carried out on the electrical system. Remember that on some vehicles, particularly those of American specification, the airbag controller stores electricity to activate the unit for up to 30 minutes after disconnection of the supply.

No work should be done on the electrical system when someone is working inside the car unless the airbags have been disconnected. They are designed to counteract against the very heavy weight of a body moving forwards in collision conditions. Expanding suddenly onto a static person could cause them injury.

Airbag units that have not activated and are to be replaced should, ideally, be set off in a controlled way before disposal. This is the procedure:

- a separate feed must be provided from a switch and a battery placed outside and away from the vehicle,
- the vehicle doors should be closed, personnel must be cleared from the vicinity,
- operate the switch.

If the airbag does not activate, allow some time to elapse before checking the connections and closing the switch

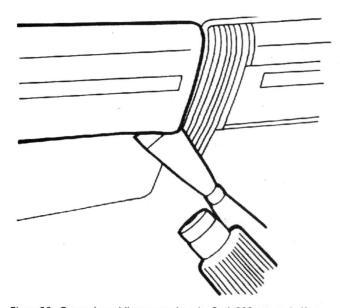

Figure 26 External moulding removal on the Saab 900; a putty knife covered with adhesive tape is used to part the moulding from the adhesive tape and release the clips; the bottom of the moulding is gently heated with a hot-air gun

again. Once activated the airbag module (see Fig. 29) may be placed into normal scrap. An inactivated unit must be handled by a specialist controlled waste contractor because the chemical content can be very toxic.

The gas produced by an airbag unit is quite harmless. It is mostly nitrogen to fill the airbag and oxygen to support combustion. Nitrogen is, of course, a major part of the air that we breathe. Although the gas is filtered, some dust can be seen after an airbag has activated. This is talcum powder used to lubricate the bag itself and is harmless.

Some cars also have gas making devices to tension the seatbelts in a collision. These work in a similar way to airbags and are triggered by the airbag control unit but the chemical is often not as toxic as that used in airbags. To be on the safe side, disconnect the electrical cables before removing seatbelts which have these units. Wear gloves and a dust mask when removing spent units.

Airbag modules and gas operated seatbelt tensioners are classified as pyrotechnics (as are fireworks and flares) and undamaged units for refitting must be stored under controlled conditions. This means storing the unit in a locked room that is not a normal workplace. Your bodyshop will also have a licence to store them.

Security systems can be a source of difficulty. Where a specialist system has been installed, the installer should ideally be asked to deal with it. If the system is in working order and the guidelines for electrical dismantling are followed, there should be no problems in disconnection and reconnection later.

Air conditioning is also a system where the services of a specialist may be needed. Special equipment is used to evacuate the gas from the system. This is necessary for two reasons. Until quite recently the standard refrigerant was R12, a gas which can damage the environment and may also be very toxic. Many air conditioner systems manufactured before 1995 can only run on this material and, as it is no longer made, must be captured and processed if expensive modifications are to be avoided. There are two important safety hazards to be noted:

- This gas is heavier than air and will flow to the lowest point. If that is a pit in which someone is working they may well be asphyxiated.
- If the gas is drawn through a flame, as with someone smoking, deadly mustard gas is produced.

R12 is also claimed to be very damaging to the environment and to the ozone layer in particular. The latest refrigerant, R134a, does not have these deadly drawbacks, except to plants. Should any escape, any plants nearby will be damaged.

With both systems thick gloves should be worn to protect the skin from the cold. Welding gloves may be suitable or any made specifically to provide insulation. The workplace should be well ventilated at a low level.

When removing any electrical components, make a note of connections or label the cables or connectors (see Fig. 30). If multipin connectors are used, check that they can only be fitted one way round and in one place. Some makes and models use one colour for their wiring and careful marking is essential unless the connectors may only be fitted in one position. Even if wiring diagrams are available, reconnecting cables is much easier if the connections are clearly marked. Do remember that what may seem all too easy at the time of dismantling could be a difficult puzzle many days and many jobs later.

Removing mechanically mounted glass

I have used this term to describe glass that is held in place by means other than bonding. Traditionally this usually utilises an 'H' section flexible moulding, the glass being retained in one channel and the bodywork panel in the other.

Successful removal depends upon applying a steady pressure over as large an area as possible close to a long side. Large panels such as windscreens are best removed by two people, one to push out and the other to control the glass. One method often used is for the 'pusher' to sit on the seats and use their feet to apply pressure. Seat protection is essential and the soles of your footwear must be covered with a clean, soft cloth to avoid scratches on the glass. The soles of your shoes are not likely to scratch, but the grit in their surface certainly will. On the outside the bonnet and wings should be protected (if they are undamaged) from accidental scratching (see Fig. 31).

Some bonded glass comes in this category because it is attached to a metal frame and the glass and frame can then be unbolted as a unit. In any circumstance where many nuts are used to retain a component, as is likely here, they should all be slackened at first, a little at a time, until they are free of tension. This avoids any distortion which might otherwise occur. On some vehicles rivets must be drilled out. Use a

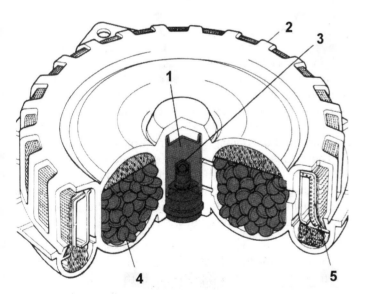

Figure 29   This is a typical airbag gas generator:
1 Priming charge
2 Outer casing
3 Igniter and detonator
4 Solid propellant
5 Metal filler
Such a unit can fill an airbag with clean, cooled nitrogen in 30 milliseconds
THE UNBURNT PROPELLANT IS EXTREMELY DANGEROUS. UNACTIVATED UNITS MUST BE DELIBERATELY ACTIVATED UNDER CONTROLLED CONDITIONS OR TAKEN AWAY BY A SPECIALIST DISPOSAL COMPANY

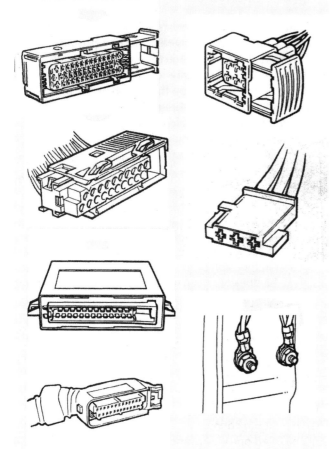

Figure 30 A variety of electrical connectors and connections; most have locking facilities, and damage or contamination of any sort is unacceptable

J76-1028

Figure 31 When undamaged body components are being removed care must be exercised to prevent damage; Jaguar require that door edges be protected with tape before removal, as circled here

double layer of heavy masking to protect the glass and the paintwork against any slip of the drill.

Many jobs will have some broken glass, either attached to the body as in the case of bonded glass, or lying as debris inside the vehicle. Some of that broken glass may have also fallen into ventilation ducting. Mechanically mounted glass has been mostly toughened, the sort that breaks into many small pieces with an impact or tension. Laminated glass is more often used now and is essential where it is bonded as part of the structure. As the name implies, this is of a sandwich structure. Laminated glass is more reluctant to break

and even after a heavy blow will usually remain largely intact. It can, however, produce very sharp slivers of glass.

Some tough leather gloves are essential when handling broken glass and it is also wise to wear eye protection. There will always be small particles when glass has been broken and these can inflict serious eye injuries. Other people in your company who will work on the vehicle and the customer must be also be considered. Particles of broken glass will almost certainly enter the ventilation ducting when a front screen is broken. These pieces must all be removed if the risk of injury from particles blown out, perhaps directly at the face, is to be avoided.

Removing bonded glass

The removal of bonded glass and its installation has always been a stressful activity, in more ways than one! So, why do manufacturers persist in using what seems to us to be a bothersome method?

It has everything to do with sales, of course, and nothing to do with ease of service. Owners and conservationists expect better fuel economy, which means that vehicles must be made lighter. One way of achieving this is to use the glass as part of the load bearing structure of the vehicle. You may think this strange as glass has little tensile strength – it objects to bending! But it is one of the strongest materials where compressive strength is concerned. Where glass is bonded into the body structure, considerable savings can be made in the amount of material used in the bodywork itself. It also allows for a smoother shape and therefore better airflow. This also improves fuel economy even further and reduces wind noise.

Our difficulty stems from the fact that insurance companies are unwilling to pay for a new windscreen when the original was not damaged in the accident and only needs to be removed for repair work to be done. A second factor is that some bonding material hardens appreciably with age. As a result vehicle and equipment manufacturers have sought hard and long to find the best method of cutting the bonding without damaging either the glass or the bodywork.

There does not seem to be any one method that can be claimed to be the most successful in all circumstances. Much also depends upon your skill, how well you know the equipment and how often you use it. There are, however, some general guidelines that will help you:

- Preparation must be thorough. Make sure that the best access to the bonding is available to you. Remove any trim where this is likely to obstruct the equipment.
- Mask-off the paintwork around the outside and inside if necessary, particularly areas which are not covered by trim.
- Place additional protection on the front wings, across the scuttle panel and on the dash.
- The equipment that you use must be in good working order.
- Where cheese-wire type cutters are used, the wire must be sound and unkinked. If winches are used they should operate smoothly and the anchoring devices work properly (see Fig. 32).

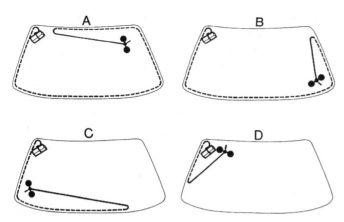

Figure 32 A typical cutting sequence when using a wire and a winch to cut windscreen bonding; the other end of the wire can be anchored or a wood block used, as here for a Saab

■ Tools using cutters must not themselves be defective and the cutters should be both the correct shape for the vehicle and sharp (see Fig. 33).

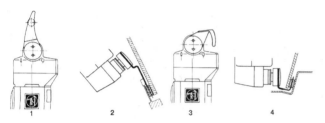

Figure 33 Shaped cutting blades for an oscillating sealing cutter; these blades are for the removal of a rear window on a Vauxhall estate – note the roller to give precise cutting depth on the straight blade in 1 and 2

■ Heating devices must reach their proper working temperature and, if incorporated into a cutting tool, the shape of the blade and its condition must be correct. If an air supply is needed make sure that there is sufficient pressure.

■ Front windscreen removal does not normally involve any risk of the glass falling free. In the case of tailgate and side windows this is quite likely. Ask a workmate to assist you.

After removal the remaining bonding must be cut away ready for any subsequent work. A sharp, chisel-shaped blade on an oscillating cutter is probably the quickest way of doing this job. A sharp knife or even a carpenter's chisel may be used instead.

Removing bolted-on panels

Bolted panels are usually confined to the front wings and perhaps metal valances (see Fig. 34). Doors and tailgates may also be attached by bolted-on hinges (see Figs 35 and 36). Some vehicles also have doors where the components, including the door skin, are bolted together.

The golden rule when stripping out an unfamiliar vehicle, particularly where little or no detailed information is available, is to take note of the order of assembly as components are removed. Work out why unusual assemblies are put

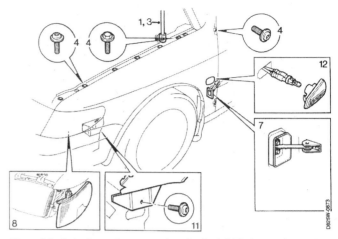

Figure 34 Bolted-on wing removal on the Saab 900; the numbers refer to the text in the manual

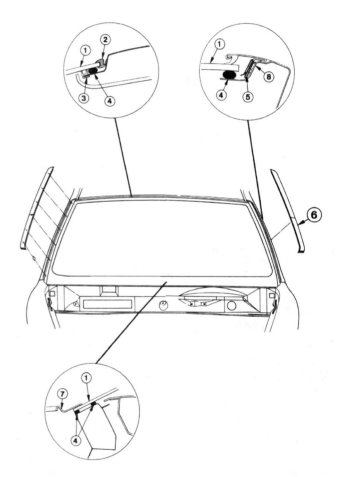

Source: Reproduced by kind permission of the Ford Motor Company Ltd.
Figure 35 Tailgate glass bonding and trims on the Ford Mondeo – note the double line of bonding on the lower edge; the numbers refer to the text in the manual

together in a particular way and tie on a label with important notes. Where adjustable items are likely to be assembled in the same position, mark the position in slots or measure the height of stops. Keep this information safely; it can save hours of work. It may also be possible to hold different sizes of bolts or screws into place with masking tape or you can screw them back into position. If neither of these options is possible, push them into a piece of old cardboard carton, marked with a simple location diagram. All this sounds as if it

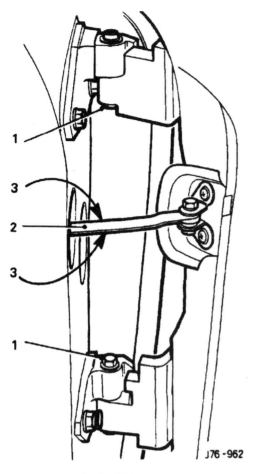

J76-962

1 Locking screw
2 Check arm
3 Grease

Figure 36 Door mounting details for a Jaguar; Locking fluid is used to prevent loosening and threads should be cleaned by tapping out

takes a long time but it is nothing compared to the time that can be wasted using trial and error when rebuilding and adjusting. At the very least keep all the components of an assembly together.

Many present day vehicles are given sophisticated anti-corrosion treatments during manufacture. These are essential if the anti-perforation warranties now commonplace are to be effective. Vinyl or other plastic based materials are often sprayed under wings and must be cut through to gain access to the fastenings and release the component. Heat is used to soften the material and a scraper or sharp knife to cut it. Take note, too, of any sacrificial anodes fitted under wing panels, under door hinges and in bolted-up door construction. These too are an essential feature of anti-corrosion protection. As in the case of boats and ships, the zinc anodes are designed to corrode away instead of the structure of the vehicle. They will certainly need renewing if they are some years old.

Removing welded panels and components

Both pneumatic and electric powered tools are suitable for this work. In your own interest you must observe the precautions outlined under Health and Safety at Work and PPE on the use of pressurised gases and mains electricity. The vehicle under repair should be protected both externally and internally. Fireproof blankets should be used to prevent damage to the interior from hot particles. Other vehicles nearby must also be protected from the effects of hot metal particles. These can become lodged in the paintfilm and eventually cause corrosion or even become embedded in the glass.

Always plan the job in front of you by looking up the recommended method in either the car makers' information or MIRRC data. If neither is available, you may have to rely solely on the replacement components to guide your choice of cutting lines and separation points. In any event, it is essential to use part panels to mark off cutting lines to suit those supplied. If you are using information, whatever the source, always check that the replacement panels conform to the suggested repair method. Car makers modify replacement panels from time to time. In such cases ask your supplier for a copy of any modification information. Be particularly watchful of the need to fix brackets or supplementary panels in place before mounting the main panel.

Cut panels with a power saw or oscillating saw. Only use a mechanical chisel in the most exceptional circumstances and certainly never do so when the edge being cut will be used as one side of a joint. The pneumatic or electrical saw must itself must be in good order, of course, fitted with a sharp and suitable blade. If the tool is pneumatic and the air pressure is low, try to overcome this problem. Before cutting check that wiring, pipes and other components are clear of the blade. Hold the tool firmly and feed it steadily along the chosen cutting line, allowing the blade to cut without applying unnecessary pressure.

Where complex curves and construction make cutting difficult, it may be necessary to make a secondary cut to gain access to the required line. This can often make your work easier where a combination of processes are used on complex joins, such as those that can be found around door shuts or on rear quarter panels, where MIG/MAG welded or brazed joints may also have to be separated.

Heavier panels, with clear space behind them, can be cut quickly and accurately with a plasma-arc cutter. An electrode creates an arc with the material which becomes molten and is blown clear with gas or compressed air. As most light vehicle repairs involve small sections or the close proximity of other components, many of them flammable, it is not generally used.

Spotwelds are best separated with a clamped-on drilling unit (see Fig. 37a), fitted with a sharp drill bit, which has a stop to give precise depth control. Drill freehand where access is difficult but check depth frequently to avoid drilling through the lower panel.

Seamwelds and brazing must be ground away to the edge with an angle grinder, or by a belt sander in tight corners (see Fig. 37b). A sharp chisel can then be used as a wedge to break any remaining join and separate the panels.

The aim in all separation work is to achieve suitable edges and joins which need little or no preparation for reassembly and closely match those of the replacement component.

Rules for panel removal:

■ Try to find information about the job,
■ Check the replacement components,

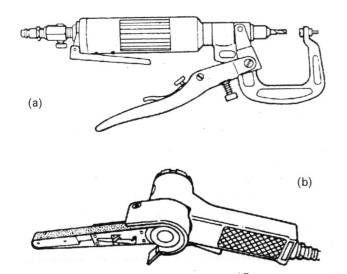

(a)

(b)

Figure 37   (a) A spot weld drilling tool and (b) a belt sander are invaluable tools for dismantling body components

- Decide what to do,
- Decide the sequence of work,
- Assemble your toolkit,
- Protect the job,
- Protect vehicles nearby,
- Protect others,
- Protect yourself
- Maintain the thickness of all retained material,
- Keep retained material undamaged,
- Think about rebuilding.

Removing bonded, clenched or riveted panels

Bonding is employed by some manufacturers for long joins which can be easily clamped up such as around a wheel arch. The technique is usually to cut the bulk of the panel away, leaving a strip of bonded material to be separated by a sharp chisel used as a wedge.

Special techniques and tooling have been developed to handle the newer methods of fastening in use on some aluminium bodies. It is unlikely that you will be asked to repair such vehicles unless the bodyshop is an approved repairer. In this case the tooling will be available and training given in the necessary techniques.

Assembling Components

Health and safety at work and PPE

Some of the general guidelines given at the beginning of the chapter also apply to assembly work. Specific guidance is given under the subjects and in the table below:

Before starting any job check that your own PPE (personal protective equipment) is in good order, that you have sufficient disposable items and that you are aware of any special requirements for any particular activity.

Sources of information

The primary source of vehicle information should be the manufacturer or that issued by the MIRRC (Thatcham).

Hazard	Protection
Sharp metal	Riggers' gloves Industrial overalls. Footwear to BS 1870 (200 joules) with steel midsole.
Flying particles.	Goggles or face visor to BS2092 Grade 2 or EN 166 grade 'F'.
Noise of grinding, cutting or drilling equipment	Ear defenders to BS 6344 Part 1 Warnings to others in the vicinity or noise screening.
MIG/MAG welding flash.	Suitable tinted visor to BS 676 or automatic adjusting visor.
Hand and body burns from hot metal.	Leather welding gloves. Flame retardent welding overalls.
Pressurised gases such as compressed air can be extremely dangerous if used in an uncontrolled manner.	Compressed air and gases must only be used with equipment that provides full shut-off and directional control. Normal workwear provides no protection against pressurised gases.
Isocyanates in some bonding agents.	There is a low level of isocyanate in some bonding adhesives; this will only be a problem to someone sensitised to isocyanate, who should wear a suitable respirator.
Noxious fumes from welding	Face masks to BS 6016 – use Grade 3 for MIG/MAG welding to protect from ozone (all metals) and zinc oxides (galvanised steel); or use EN 149 class FFP 2S
Paint stripper. Oxalic acid.	Full protection for handling acids including face visor, respirator, gloves, apron and boots.

Tool and equipment instructions may give information specific to vehicle types.

Consumables will have general instructions which are adequate in most instances. Where there is a car maker's body warranty in force the car maker's own branded products or alternatives acceptable to them should be used. In the case of insurance based repairs the insurers' written instructions should be followed as any subsequent problems not due to poor workmanship or wrong procedures will become their responsibility.

Equipment and hand tools

Every job should start with considering how the work should be done, and then deciding which equipment and tools will be needed to complete it successfully. Make sure that your own tools are in good condition and to hand. Check that any special tools are available for your use. This is particularly important where items such as jig brackets are hired-in for a specific job. Now is the time to prepare items such as the electrodes of spot welders. Select all the versions that you will need and check them for wear. Rectify them as necessary. Details of the correct treatment and adjustment of welding equipment are in the relevant welding sections.

Vehicle and component preparation

Preparation is in two main stages; the vehicle itself and the components which are to be attached. The methods of attachment are either mechanical or by a fusing or bonding. These latter methods impose their own special needs.

Where welded or brazed components have been carefully removed with the correct tools, the flanges and attachment points will only need a light dressing. Excess brazing can be cut back if necessary with a belt sander (see Fig. 37b). Bonded areas will need the bonding agent removed. This can usually be achieved by using a rotary wire brush. For aluminium the wire brush should be of stainless steel. In all these activities PPE should include eye guards to BS 2092 Grade 2 or EN 166 grade 'F', to provide adequate protection against flying metal particles. A final dressing with an angle grinder, or a belt sander for areas of difficult access, may be given provided that only the remnants of any removed component or attachment residues are skimmed off. This is especially important for vehicles with galvanised body panels. Always try to keep such coatings undamaged if at all possible. It is even more important that the vehicle panel or component is not reduced in thickness so that the body structure retains its strength. However, cleanliness is essential for the successful attachment of components by welding, brazing or bonding. Cleanliness is also of the utmost importance where box sections such as sills have been cut open. These are often treated with wax preservative, to which swarf and other debris can adhere. Ferrous particles trapped in the preservative could cause corrosion in time. Cover the opening or fill it with paper or cloth to prevent contamination, taking care that the debris does not fall into the opening when you remove the protection.

In the case of mechanically fitted components, such as bolted-on wings, vinyl or other sealers may be cut away with

a sharp knife or even a wood chisel. Time can be saved by using an oscillating chisel cutter made for the removal of glass bonding. All these tools must be kept sharp for ease and speed of working and a sharpening stone is invaluable. A rotary wire brush or mule skinner is usually all that is needed to clean off any remaining sealer.

Body components vary greatly from make to make in the preparation that is needed. You may need to make good transit damage, clean off preservative and prepare the joins. Transit damage that may be rectified is generally confined to bent flanges or small dents. The flanges or joining edges are likely to be dressed anyway and small dents are treated in the normal fashion. If a preservative coating is covering the component, remove it with a solvent. You must, of course, wear the appropriate mask to avoid inhaling toxic fumes. If there is any doubt regarding the nature of the surface treatment, ask your supplier. Often the factory applied, anti-corrosion coatings are sanded away in ignorance of their true purpose. These may include phosphate and the 'E' coat (electrophoretic coating or electroprimer). Apart from the waste of time and effort, this may negate the vehicle maker's anti-corrosion warranty.

When the component is clean it is offered into position and any cutting or shaping necessary for a good fit is carried out (see Fig. 38). At this stage pre-drilling or punching of the flanges can be carried out where MIG/MAG plug welding is required. If it is necessary to remove pre-treatments, such as electrostatic coatings, follow the vehicle makers' guidelines. In the case of galvanised panels in particular coatings may need to be removed by paint stripper to avoid damage to the galvanised surface.

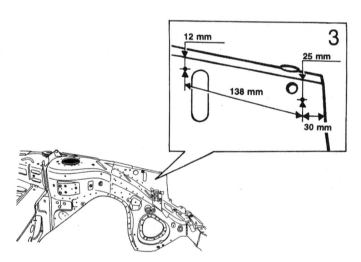

Figure 38 New components often need some modifications as in this example from the Saab 900 manual

Some vehicle makers insist that zinc based coatings are applied to flanges, particularly those for spot welding. Such coatings are critical for galvanised bodywork and for those makes with long term anti-corrosion warranties who require it. The coating should, of course, be acceptable to the maker concerned and it is always best to use that which they supply or recommend. The use of weld through zinc primers is, in any event, good practice for any steel bodywork as it will do much to prevent corrosion within the flange (see Fig. 39).

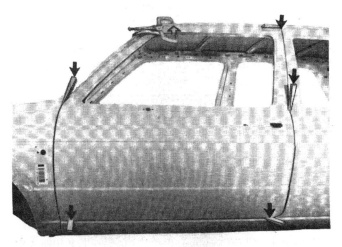

Figure 39 (a) Clamping components together is a familiar routine in the panelshop (b) Doors with welded-on hinges need great care in aligning and fixing the door in position; this procedure is from a Volkswagen manual

There is a tendency in non-franchised bodyshops to ignore such requirements, particularly where insurance funded repairs are carried out. But do remember that at least one vehicle manufacturer has an anti-corrosion warranty that lasts for ten years! The repaired part of the car can become the responsibility of the repairer, if the repairs are not carried out properly and corrosion is subsequently found in the area of the repair. Where an owner or an insurance company insist on different repair techniques, such instructions must be obtained in writing and kept for the life of the warranty.

Part finishing components before assembly is sometimes a speedy and acceptable alternative to painting everything after repair. One manufacturer asks for the inside of panels that are visible, such as a rear quarter panel, to be painted before welding into place. This same maker only insists on primer and topcoat in such locations. Where components are mechanically fastened they may be totally finished before assembly, as in the case of bumpers and bolted-on wings. Check the fit and assembly before the finishing process, especially front wings. Sometimes brackets may have to be welded into place and holes cut or filled.

The majority of vehicles today are equipped with systems using electronic control. The controllers, or 'black boxes' as they are often called, are miniature computers and usually very expensive to replace. Incorporating the vehicle into an external electrical circuit, as happens in any form of electrical welding, may result in surges of current flowing throughout the vehicle and this can destroy the electronic circuitry. Each black box should be disconnected from its circuit by unplugging the connector block or blocks. The greatest care must be taken of both the connector blocks and the black boxes, particularly the cleanliness and condition of the connection terminals. The connectors on the vehicle should be wrapped in polythene and the black boxes removed and stored carefully. Don't forget to make a note of their locations if any of the connectors are identical to another.

Body structure materials

Until relatively recently virtually all mass produced, light vehicle bodywork has been made from good quality low-carbon steel of various grades and thicknesses. It is a material of high tensile strength and ductility. These properties are essential to the effectiveness of the energy absorbing crumple zones to be found on many vehicles. The bodywork can also be easily fastened together by resistance spot welding, a process that is well suited to automated construction, without the loss of these qualities. Coating mild steel with zinc as an anti-corrosion treatment by the process called galvanising is now quite common. Pressings can be made from pre-coated steel, coated after pressing or even after assembly. Zinc may also be applied in powder form by some manufacturers.

The search for better fuel economy and so for lighter vehicles has resulted in the use of high-strength, low-alloy steels (HSLA), often referred to as high-strength steels (HSS). There is renewed interest too in aluminium. Cars are already available in the specialist or executive class and two major American based manufacturers are testing new models made in aluminium alloy. Sometimes vehicles made from these materials are built in different ways; special tooling and training may also be needed before you can carry out successful repairs in the bodyshop.

Here are some important points about the welding of bodywork constructed in these materials.

Conventional low-carbon steel needs no more special treatment than that which is usual in the normal repair process.

Galvanised coatings create no greater need than slightly higher currents and electrode pressure when spot welding. MIG/MAG welding calls for protection for yourself from the zinc oxide released when welding with an arc and careful set up and gun control to avoid spattering.

HSLA (HSS) steels are another matter altogether. First, they vary a great deal in their composition so it is not always easy to say, with any certainty, precisely what effect welding will have on them. One thing is certain. Excessive heat will change their ductility quite dramatically and they will become either soft or brittle. This could have disastrous consequences during day-to-day use of the vehicle or in a future accident. Resistance spot welding with its very local heating and low temperatures is ideal. The use of any form of flame or arc welding will almost certainly result in changes to the material. Oxy-acetylene welding must never be used. The only possible exception is that of carefully controlled

MIG/MAG plug welding. Provided that the heat build-up is kept very localised this method may be satisfactory. One other concern is that the wire introduced into the weld as part of the MIG/MAG welding process will almost certainly be substantially weaker in composition than the HSS which it is joining together.

Aluminium presents challenges, too. MIG/MAG and TIG welding are quite suitable although special training is needed for successful TIG work. Resistance spot welding is possible but special equipment is needed. A recent design of an all-aluminium bodied car employs riveting and clinching during production in preference to spot welding (see Fig. 40). Repair techniques for minor damage to this vehicle only involves the use of bonded, bolted-on and riveted assemblies.

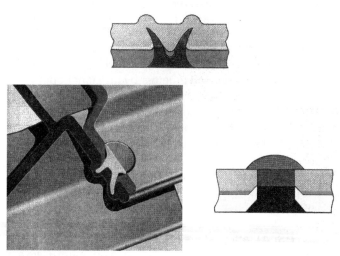

Figure 40 On the all-aluminium Audi A8 rivets are used instead of spot welds; here you see sectioned views of the rivet used in production and, on the right, the type used in repair

Attaching components by resistance spot welding

This is the most common way of joining sheet metal components together. A present day steel monocoque car body may have as many as 6 000 spot welds. Spot welding provides a strong joint between components provided that it is carried out correctly. To do that some guidelines need to be followed which, if you know the reasons for them, are common sense. Remember that the effectiveness of the safety measures designed into the vehicle are often almost totally dependant on how well the spot welds are made at the critical parts of the body.

A spot weld is a fusion of two or more pieces of metal, in a circular shape, about 4 millimetres across. The metal is made molten by the passage of a heavy current through the tips of 2 electrodes which have the components to be joined clamped between them. Clamping pressure must be at least 200 pounds. The voltage is very low, about 3 volts to 5 volts, with an amperage of between 5 000 amps and 10 000 amps. This is substantially lower than that used in production where the current may reach 80 000 amps. It is the metal resisting the passage of such a large amount of electrical power that causes it to become hot. The temperature reached is around 1 300 degrees Celsius. The metal between the electrodes

melts and the panels fuse together. The little button of fused metal is called the nugget. A properly made spot weld is ductile; that is to say it is not brittle and will not easily fracture.

Good preparation, as always, is the secret of successful spot welds. The flanges to be joined must be clean and able to fit together neatly. Use body clamps to hold the panels in place. Galvanised panels should have paint or sealer removed. Every effort should be made to retain the galvanised zinc coating. This will melt and form a protective ring around the weld nugget. The best welding equipment melts the zinc before the burst of welding current is applied. Many car makers ask you to use a welding primer containing zinc. If the car is still under a body warranty the makers may demand that you only use an approved primer. It is good practice to use a zinc primer on all spot welded joints. The welding equipment must be able to give the slightly higher current needed for zinc coated materials and all the other conditions must be met.

Careful cutting, cleaning and dressing of the parts to be joined is essential. Trying to take short cuts in preparation and neglecting to set-up the equipment correctly will result in difficulties, an inadequate job or both.

Your equipment must be of good quality, in sound condition and properly maintained. Electrode tips must be dressed to a maximum of 4 mm in diameter to concentrate the current flow. Electrodes that look like mushrooms may give a large contact area but that spreads the current flow and so a sound spot weld is impossible. A poor fit and loose clamping of the electrodes into the arms may also result in heavy current loss. This can be caused by a small contact area through either poor design in the first place, through wear or careless fitting. The larger the area of contact in the arms, the greater the current flow and the likelihood of a good weld. The electrodes themselves must be exactly opposite each other when clamped into position (see Fig. 41). These factors apply to all spot welding units although it is often critical with those where the transformer is mounted on the arms. The weight of these units limits the size of the transformer and thus the current available for welding.

Thorough preparation of both equipment and the repair is the only way to obtain good spot welds. Only then will the vehicle bodywork function correctly in normal use and protect the occupants in accidents.

Here is your checklist for successful spot welding.

- The faces of the metal should be clean to enable a full flow of current to develop.
- They must also be pressed together where the weld is to take place for the same reason.
- The electrodes must be exactly opposite each other.
- The area of the electrode tips is critical; it must never be more than 4 millimetres in diameter.
- There must be a good area of contact between the electrodes and the arms.
- Pressure at the electrode tips should be in excess of 200 pounds. When spot welding galvanised materials, the highest possible tip pressure should be used.
- The electrodes must be sufficiently far away from the nearest weld to prevent the current taking the easiest route and back tracking through the one nearby. One manufacturer quotes a minimum figure of 13mm.

Figure 41   The electrodes of resistance spot welders must be perfectly aligned and not more than 4mm in diameter at the tip

- Equally, however, there must be sufficient welds to obtain the necessary strength.
- Use the same number of welds as in the original construction. These may be given in the car maker's service literature. They are always to be found in Thatcham data sheets.
- Best practice requires at least one test strip to be welded from material cut from the repair, or its exact equivalent, to test the equipment and the weld strength.
- The trial spot weld should be tested by pulling one strip vertically away from the other. The weld is successful if one strip 'unbuttons' from the other, leaving the spot weld nugget in place on one of the test strips and a hole in the other (see Fig. 42).

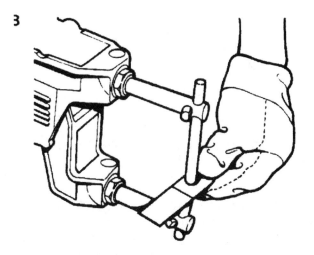

Figure 42   A test strip to check the welder settings on every job is obligatory in aircraft repair; it should be routine for every car job, too

- One sided spot welding should ideally only be used for cosmetic panels.
- In one sided spot welding the sheets to be joined must be in good contact. The only pressure available at the electrode is that resulting from the effort of the operator.
- In one sided spot welding the earth connection or connections should be as close to the welding as possible to minimise current loss.

Attaching components by MIG/MAG and TIG welding

Health and safety precautions consist of protection against hot metal, toxic gases and the bright flash of the welding arc (see Fig. 43). Protective clothing consisting of flame proof welding overalls and heat resisting gloves, usually heavy leather, should be worn. Eye protection is by conventional welder's visor or hand shield with either permanently tinted glass of suitable type or the newer automatic dimming visor. During the welding of galvanised material, potentially dangerous zinc oxide is given off and a suitable mask should be used. Powered respirators are now available that clean the air as it is drawn into the mask. As with any absorbent type filter, the replaceable element must be changed in good time if gas inhalation is to be avoided. Alternatively, an air-fed face mask may be used, provided that it meets all other criteria.

You should also remember that some of the gases are heavier than air. If welding is carried out in an enclosed space which is not vented at the bottom, a build-up of suffocating gases can occur.

The protection of others in the workplace, or those who may enter it, must also be considered. Whilst an occasional exposure to the flash of arc welding, known as 'arc-eye', usually results in only temporary discomfort, precautions should be taken to prevent it. The easiest way is to draw welding screens around the job. Fume extraction, either portable or built-in, should be used at all times.

MIG (metal inert gas) welding (see Fig. 44), one of the methods known to welders as MAGS welding (metal arc gas-shielded), is widely used in the repair of low carbon steel bodywork. When used with an active gas it is called MAG (metal active gas) welding. MIG/MAG welding equipment is also available with electronic control to cope with the more complex needs of aluminium welding. It enables panel technicians who have been given a little training in its use to achieve weld quality virtually as good as that produced by a skilled TIG operator.

TIG (tungsten inert gas) welding (see Fig. 45), also known as TAGS (tungsten arc gas-shielded), is an alternative used in some body repair shops. It is often favoured by those renovating specialist cars and for use on aluminium bodywork. TIG is particularly good at producing high-quality welds in thin materials when used by a skilled operator.

Both systems employ the creation of an electrical arc between the welding electrode and the work, which is attached to a welding current return. The arcing causes a small pool of molten metal to form. When a welding rod or wire is entered into the pool it melts and causes a build-up of metal. Two panels can be joined together by moving the torch along the seam, feeding a welding rod into the pool at the same time. The difference between the two systems is that in TIG welding a permanent tungsten electrode which does not melt is used. The filler metal is fed into the molten metal pool by hand. MIG/MAG welding employs a continuously feeding metal wire electrode, which is melted into the join automatically.

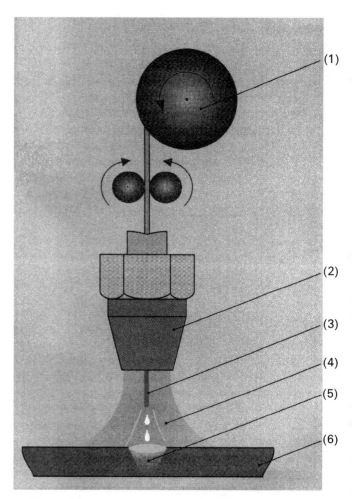

Figure 44 This diagram of MIG/MAG welding shows the rollers drawing the welding wire from the reel (1) and feeding it through the nozzle (2) where it becomes the feedwire and electrode combined (3); oxygen is prevented from reaching the molten metal (5) between the work pieces (6) by a shield of gas (4)

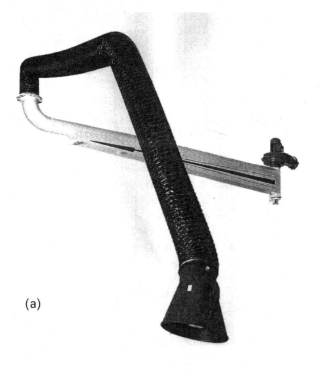

(a)

(b)

Figure 43   (a) An arm mounted, welding fume extractor and (b) its active carbon filter

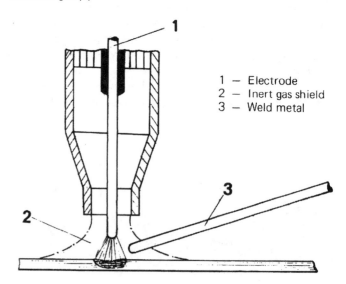

1 — Electrode
2 — Inert gas shield
3 — Weld metal

Figure 45   In TIG welding there is a permanent electrode and the weld metal is introduced through the gas shield

Both types employ a screen of gas around the electrode, the arc and the pool of molten metal to exclude air. This gas is either carbon dioxide (CO_2) or more usually a mix of carbon dioxide and argon. Helium may also be added and is particularly useful for brazing. Almost pure argon is used for aluminium welding.

The wire used in MIG/MAG machines will form part of the body structure. Good quality, layer wound wire is essential for best quality welds and for smooth feeding. Aluminium vehicles may need special wire with higher than usual silicon content.

The workpiece forms part of an electrical circuit, just as it does with resistance spot welding. A higher voltage is used, between 15 volts and 25 volts, but a much lower current at between 5 amps and 200 amps. Good electrical contact between the components is desirable for good quality welding.

Here are some important guidelines for successful MIG/MAG and TIG welding:

■ The shielding gas must be suitable for the metals being welded.
■ The wire or filler rod must be of good quality and of the correct metal.
■ To ensure a smooth feed, the welding wire should be layer wound.
■ Metalwork must be clean; aluminium should be de-oxidised with a stainless steel wire brush.
■ Adjust the current to match the size of electrode being used.
■ With galvanised panels slow the wire feed rate a little.
■ When welding galvanised materials the torch should be held at an angle of between 80° and 90° from the panel, that is just off the perpendicular.

Each and every weld that you make should be checked for quality, particularly those at critical points. The ability to self-check your work is also essential if you are to maintain or improve your skills as a welder. So what do you look for? Here are the main visual signs of a good MIG/MAG weld:

■ The weld metal and the parent panel should be fused together along their entire length.
■ Components must be at least as thick as they were originally.
■ Butt joins must show a clear and narrow underbead throughout their entire length (see Fig. 46).
■ Weld build-up should be seen by a slightly raised surface.
■ The weld seam should be of regular pattern and free from holes, slag or burn marks (see Fig. 47).
■ Panel discolouration should be even on each side and confined as close to the weld as possible.
■ The weld and the panels should be free of spatter (see Fig. 48).

MIG/MAG plug welding

Plug welding using the MIG/MAG process is particularly useful for panels where access is restricted to one side. The panel being fitted has a series of holes punched into the welding flange with a hole punch. When the panel is clamped into position the parent panel provides a bottom to the hole. The MIG/MAG torch is held over this 'crater' and it is filled with a plug of weld. The finished surface should be level or just slightly raised.

Although more heat is generated than by resistance spot welding, plug welds carefully made are very effective and particularly suitable for joining components where high loading may be expected (see Fig. 49).

MIG/MAG plug welding may also be done using a high-powered welder fitted with a special nozzle (see Fig. 50).

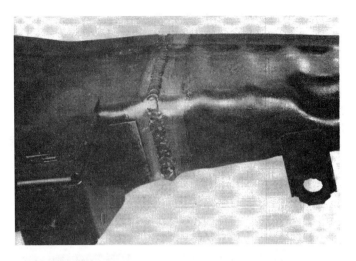

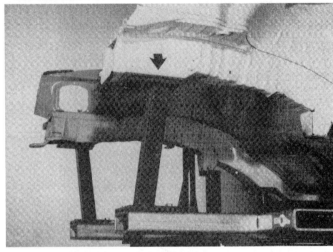

Figure 46 A MIG butt weld repair to a deformation element: strength and perfect alignment are essential

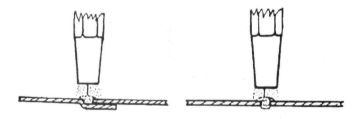

Figure 47 (a) an overlap and (b) a MIG butt welding joint

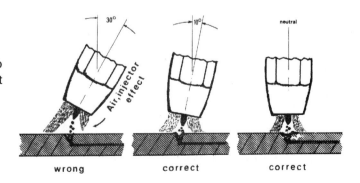

Figure 48 To avoid spatter and a porous joint when MIG welding galvanised panels, the torch must be kept within 80 and 90 degrees of the surface

FRONT END PANEL OPERATIONS

Front Chassis Leg Assembly Section (B):
In replacement, trim and discard the rear section of the new chassis leg assembly to form a seam welded butt joint with the existing panel at **2** in **Fig A**. Make a further seam weld at the centre section of **3** as there is limited welding access in this area.

Punch and drill holes for plug welding in the new panel at **1**, **3** and **4** as there is no spot welding access in these areas. Make further plug welds at **5** using the holes left by the spot weld cutter.

Fig. A

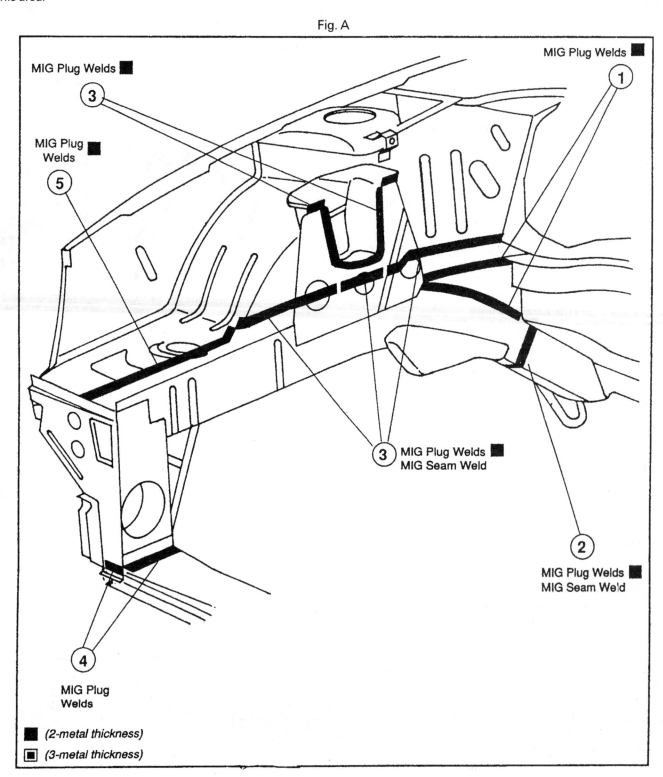

MIG Plug Welds ■

MIG Plug Welds ■

③

MIG Plug Welds ■

⑤

①

③ MIG Plug Welds ■
MIG Seam Weld

②
MIG Plug Welds ■
MIG Seam Weld

④
MIG Plug Welds

■ *(2-metal thickness)*
▣ *(3-metal thickness)*

Figure 49 MIRRC Methods manual information: welding diagram

**FORD MONDEO
4- AND 5-DOOR**

PANEL, PAINT & M.E.T. OPERATIONS
REAR END REPAIRS

REAR WING METHOD 'A' AND REAR CORNER REINFORCEMENT IN COMBINATION (4-DOOR)

This panel repair operation details replacement of a rear wing cut to fit the existing panel between dogleg and wheel-arch (see Section A page A3 Fig. 14 refs. 025(1) and 209).

PANEL:

REMOVAL

Cut out the spot welds at the rear section of the weatherseal retainer at **No. 12** in **Fig. 33** and bend it aside for access to the upper quarter, then cut the rear wing at points **J, K** and **L**. Remove the rear corner reinforcement **No. 5** and cut out the remaining spot welds. Remove panel bulk and metal remnants, noting the 3-metal thickness at the screen panel **No. 3**.

Fig. A

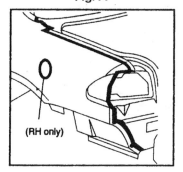

(RH only)

REPLACEMENT

Drill and punch holes in the new rear wing for plug welding at **Nos. 3, 4** and **6** as there is a lack of spot welding access at these points, also at the rear corner reinforcement **No. 5** due to excessive metal thickness. Make further plug welds at Nos. **3, 11** and **12** using the holes left by the spot weld cutter

Transfer the striker reinforcement and bumper side mounting to the new panel at **No. 10**, plug welding the striker reinforcement into position using the holes left by the spot weld cutter. Apply metal-to-metal adhesive to these items before offering up as shown in **Fig. 33** (see also **Introduction** section **page 10**), and also at the outer wheel arch.

Fit new anti-drum pads to panel interior before painting operations.

PAINT:

Apply joint sealant to new panel as shown in **Fig. A** after primer application and before colour coats (see also **Introduction** section **page 10**).

WELDING TABLE

No.	Location	Remove Factory Joint	Replace Repair Joint
1	To existing panel at upper quarter		1 × 200mm MIG seam weld.
2	To inner wing at rear screen aperture	12 resistance spot welds	Single row of resistance spot welds ARO arms 242A or equivalent.
3	To boot lid drain channel	18 resistance spot welds	18 MIG plug welds.
4	To tail lamp panel	9 resistance spot welds 1 × 20mm MIG tack weld	9 MIG plug welds. 1 × 20mm MIG tack weld.
5	Rear corner reinforcement to wing	7 resistance spot welds	7 MIG plug welds.
6	To rear panel	5 resistance spot welds	5 MIG plug welds.
7	To boot floor side extension	8 resistance spot welds	Single row of resistance spot welds ARO arms 242A or equivalent.
8	To outer wheelarch	15 resistance spot welds	Single row of resistance spot welds ARO arms 242A or equivalent *at rear section* and 1 25012 *at inturned flange*.
9	To existing panel at dogleg		1 × 300mm MIG seam weld.
10	Striker reinforcement and bumper side mounting to rear wing	19 resistance spot welds	4 resistance spot welds ARO arms 242A *at bumper side mounting*. 15 MIG plug welds *at striker reinforcement*.
11	To inner wing at dogleg and extractor vent aperture	14 resistance spot welds	Single row of resistance spot welds ARO arms 242A or equivalent. 2 MIG plug welds.
12	Weatherseal retainer to wing	6 resistance spot welds	6 MIG plug welds.
13	To inner wing at fuel filler aperture (RH only)	4 resistance spot welds	4 resistance spot welds ARO arms 246A or equivalent.

Figure 50 MIRRC Methods manual information: welding data

The torch is held a set distance from the panel and melts a hole in the top piece at the same time as creating the plug of weld. This method saves time but should only be used when the equipment has sufficient power (160 amps or more) and the operator has been trained in its use. It is not unknown for a panel to fall away after a series of plug welds have been made that look perfect!

Attaching components by brazing

Brazing is favoured by some manufacturers for panel joins in areas where complex shapes are brought together and a smooth appearance is required, such as around lamp clusters or in door shuts. Traditionally, the materials are joined together by running a lower melting point metal between them with the aid of a flux. The purpose of the flux is to break down the surface tension of the molten filler metal, enabling it to flow into the joint. The filler metal takes the form of brazing rods or wire and those used in the repair of steel bodywork are of brass or bronze. An aluminium-silicon rod is used for aluminium.

The mating surfaces must be clean; a rotary wire brush is generally suitable for steel. Only stainless steel or non-metallic brushes may be used on aluminium to avoid the danger of metallic corrosion from particles of the brush trapped in the surface. The mating surfaces must be close together to enable capillary action to carry the filler metal into the joint. The ideal gap is no more than 0.1 mm. As this is smaller than the thinnest human hair, the best fit that you can make between the surfaces is likely to be more, so always work to obtain the very best fit possible.

The technique involves heating the mating surfaces generally until the filler metal melts when it touches the surface. The flame of the torch must be kept moving to avoid local overheating. Slightly overfeed the joint to allow for the smallest amount of dressing back to give a smooth contour.

The flux used for brazing is corrosive. For this reason few manufacturers use the method and even then in only one or two positions where corrosion is not normally a problem. Brazing during repair should be restricted to those points where it was applied in manufacture and then used as carefully as possible. Flux must be applied sparingly and any residues thoroughly cleaned away.

A form of brazing is possible with MIG/MAG welders, although this is really welding by using a brazing wire. This may be used for cosmetic panels because a very neat joint is obtained needing little dressing. It is unsuitable for load bearing panels due to the weaker nature of the filler metal.

Attaching components by adhesive bonding

Joining components together with non-metallic materials is a 'space-age' technique now used in production and during repair. This method has been made possible by the development of 'two-pack' resin based bonding agents. In these a measured amount of a separate hardener is added to a basic mix. This starts a chemical reaction which allows only a limited amount of time to complete the bonding process. The quantity of hardener can govern the curing time and may also affect the strength of the finished bond. The tempera-

ture of the surrounding environment (the ambient temperature) and that of the components to be joined may also be important and the product directions should be followed carefully.

Bonding is used in metalwork assembly and in mounting glass to the body (see Figs 51 and 52). Glass bonding is dealt with in the section on fitting glass. In bodywork, bonding is mainly used for overlap joints where the strain tries to make one panel slide against the other. This is called 'shear stress'. In such joints a comparatively large area is available for bonding which enables the adhesive to resist the stress. You can see from this that all the area of metal in a joint should normally be prepared and coated with the adhesive.

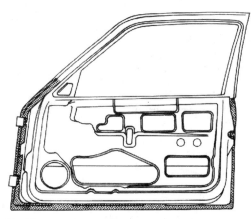

Figure 51 Door panels are often bonded; the shaded area shows the bonded section of a Saab 900 door

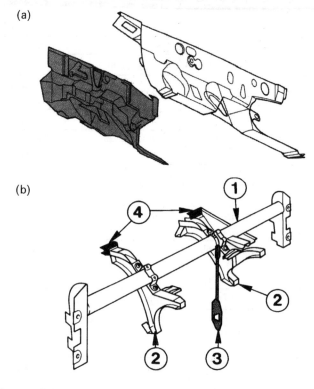

Source: Reproduced by kind permission of the Ford Motor Company Ltd.
Figure 52 Two design features of the bulkhead on the Ford Mondeo: (a) sound insulation and (b) a bolted and cross member
1 Cross member
2 Retaining elements
3 Additional mounting to floor assembly
4 Mounting surface bonded to bulkhead

Successful bonding demands a high standard of preparation of all tools, materials and the panels. Here are some guidelines that will help you:

- Flanges should be the best possible fit.
- Check carefully that the panel can be clamped into place without difficulty.
- The flanges must be absolutely clean and dry.
- Read the adhesive instructions carefully.
- Make sure that all the tools needed to complete the operation are to hand.
- Obtain a new, clean palette for mixing.
- Use a clean spatula for mixing and spreading.
- Thorough mixing of the two components is essential; watch the colour change.
- Mix and apply enough adhesive to fill the joint and provide some excess to squeeze out.
- Folding must be carried out before the adhesive has set.
- If the repair instructions also call for spot welds these must also be carried out before the adhesive has set.
- Wipe off excess carefully and dispose of the waste correctly.

Two of these actions are vital in preventing corrosion inside the joint. Mix enough adhesive to overfill the joint and then spread it in an evenly thick coating. Together these actions will help to avoid the air pockets which cause corrosion by holding moisture. It follows that any clamping or folding of the joint should also be done carefully.

Single component types of bonding material are also available and are usually to be found in the form of thread locking fluids. Single pack products of the anaerobic type cure by the exclusion of air.

Attaching components by mechanical fastening

The way in which parts are held together mechanically may be split into two groups: using threaded fasteners or clamping parts together in other ways.

Threaded fasteners have either accurate machine threads, as with bolts and nuts, or a coarse, rougher thread as can be found on self-tapping screws. Machine threaded bolts and nuts are always used on mechanical units such as the engine or on the suspension. Here the degree of tightness or torque must be very precise if the bolts are not to break and yet stay tight. That does not mean that parts should not be carefully tightened but the torque is not so exact and the stress is often less.

By long tradition normal bolts are made with a six sided, hexagon head. Nuts too are usually this shape. The hexagon provides the parallel faces needed for open-ended spanners. The evenly spaced corners allow ring spanners and sockets to be used. Another fixing is the stud. This is like a bolt with no head and a thread at each end. It is made with machine threads and one end screws into a metal housing. When the parts are assembled, the other end of the stud protrudes from the housing, onto which a nut is placed.

There are also bolts with cylindrical heads which have hexagon tightening faces cut internally. These 'socket cap screws' are sometimes given their original name of Allan screws. Nowadays, the internal drive may be of different shapes, such as in TORX® screws. This design gives large contact areas and no concentrated stress in the bolt head, as well as a positive drive. The correct type of key or socket must be used to undo or tighten these bolts or screws.

Screws usually have a circular head. This may be cylindrical, as with socket headed screws, countersunk with a flat top or domed. Machine threaded screws are normally cap screws or countersunk to provide a more solid head for positive driving. Coarse threaded screws may have any head design, depending on use. A variety of pressed steel or even plastic nuts may also be used with coarse threaded screws, particularly on trim fittings. Nylon screws are sometimes used for fastening lightly loaded items.

Bolts, nuts and screws are made in a variety of metals and, as we have seen already, even plastics. Those used on motor vehicles are generally of steel, usually with some form of coating to retard corrosion. For steel bodies this is usually phosphate, or electroplating of cadmium or zinc. Aluminium bodies are quite different and you need to understand why.

The metals used in making vehicles each have particular properties. Their colour and hardness are two ways in which they differ. They also have different electrical properties and it is this relationship which is the most important as far as corrosion is concerned. Metals touching each other can corrode by electrolytic action. The more they differ in electrical properties, the more they will corrode. Untreated steel will cause aluminium to corrode away at the point of contact. All steel bolts, screws and other fastenings for use with aluminium must be treated with an aluminium or zinc aluminium protective coating. One car-maker goes as far as colour tinting the items that may be used on the aluminium bodies of their cars.

Screwed fastenings may often be re-used, particularly on bodywork, where the stress applied to such fittings is not very high. But there are occasions when replacement may be advisable or even compulsory. The fastenings used on suspensions, for example, may be exposed to heavy loading. Some bolts used on mechanical components are designed to stretch when correctly tightened. To avoid thread or fatigue failure, all special bolts and nuts, and those used in assemblies that contribute to active safety, should be renewed. Nuts with an insert, an oval thread or even sometimes threads that are not exactly identical in pitch to the bolt, are used to avoid loosening due to vibration or stress. Most of these will need renewal after dismantling.

All the screws, bolts and studs used on a motor vehicle are made from materials that will withstand the strains of use, as well as those imposed by tightening them. Most, if not all, are high-tensile. It is very important that only the correct fastenings are used for any important component. The only way to be sure is to use genuine parts supplied by the vehicle maker.

Screwed fastenings are probably damaged more often by overtightening than by any other cause. Service information generally gives tightening torques, certainly for all the most important connections. Apart from nuts and bolts on engines or transmissions, those which must be tightened most carefully are those on the running gear. Particularly important is the humble and vitally important wheel bolt or nut. Correctly tightened, the wheel may be removed without the use of undue force. More importantly, the brake disc, drum or hub will not be distorted. By using a torque wrench that is correctly set you can be certain that the fastener is tight enough to do the job and is not damaged or causing damage by overtightening.

An important component that is needed for many bolts, nuts or screws is the washer. Shakeproof, spring or plain, they can make the difference between success or failure when putting parts together. Spring and shakeproof washers are used to prevent a screwed fastener from working loose due to vibration or strain. Plain washers are used to spread the load on softer materials or to relieve friction on bolts or nuts that must be very tight. Be guided by the maker's fitment in the first place and always renew spring and shakeproof washers.

A modern alternative for securing screwed fittings is locking fluid. These resin based compounds cure by the exclusion of air. Normal use demands only a drop of fluid on each fastener. Too much can make it almost impossible to undo the connection!

Non-threaded fasteners are used in many shapes and sizes. Plastic or nylon is often used for trim clips, both inside and outside. Remember that low temperatures will often make them brittle, especially if they are some years old. Always warm them before removal or fitting if they have been in a cold workshop overnight.

One sided (blind) rivets, or pop rivets as they are usually known, are also used for retaining trim or even glazing. There are several requirements for successful riveting:

- The components to be joined must be pressed tightly together.
- The holes in the components should be of the correct size.
- The rivet material must be correct for the application.
- The rivet must be the correct diameter and length.
- Pressure must be maintained while operating the riveting tool.
- Complete the compression of the rivet to break off the pin.

Some bodywork may call for the use of other types of rivets. At the time of writing, repair techniques for one of the new generation of aluminium cars involves bolting or bonding and riveting components together, instead of the clinching or punch riveting used in production. In such cases the correct tooling and parts must be used to re-build the car safely and to meet the needs of the warranty. Do remember that specially coated bolts, screws, washers or rivets may be needed for aluminium bodywork.

Filling and sealing

These processes are part of the NVQ syllabus for painters, but because panel technicians often do these jobs, it seems appropriate to consider some aspects of them at this point. In the normal work sequence, filling and sealing should always be separated by the application of priming coats in the spraybooth. For the purposes of this book, it is convenient to look at them one after the other.

Filling

In many bodyshops, the repaired vehicle must be presented to the painter at a stage where the final panel contours have been achieved. The painter's job is to apply the finish to the appropriate standard. The panel technician should be able to select, prepare and apply fillers that will help in achieving this standard.

Start by selecting a filler that is suitable for the job and, especially on galvanised or aluminium bodies, be sure to use those approved or supplied by the vehicle maker. Fillers made for general use must be compatible with the paint system that will be used on the vehicle. Fillers approved or supplied by a vehicle maker will always be compatible with an approved paint system. When selecting a filler from a general range, make sure that it is suitable for your purpose. Substrates such as galvanised steel and aluminium have special needs. Remember also that there may be special needs where anti-corrosion warranties are still in force.

The quality of the repair should be such that only a fine grade of filler, applied thinly, is needed for surface shaping. In any event, filler depth should never exceed 4 millimetres. A deeper indentation at, for example, a complex panel join should be body soldered (lead loaded) and some information on this technique will be found in Chapter 4.

Fillers are often of the two-pack polyester type which, provided that they are mixed and used correctly, are easy to apply and to shape. The first, thick coating is applied with a spatula. This is usually followed with a brushed or sprayed fine filler to give a high-quality surface.

Successful filling is based on good mixing. This is the secret;

- mix accurately,
- mix cleanly and
- mix carefully.

Measure the amount of each part as accurately as possible. The molecules of each of the chemical constituents must be able to team up with those of the others. Any left over will cause problems such as bleaching in the subsequent coats of paint.

The base and hardener must be mixed on a clean, non-absorbent palette. This will ensure that no resin is lost from the mix. It also reduces the risk of air bubbles becoming trapped in the filler. A clean palette makes certain that the chemical action will not be affected. Clean steel sheet is ideal. For the same reasons the spatula should be non-absorbent. If it is also slightly flexible, spreading and shaping the filler will be easier.

Mixing should be done in a way that spreads the two parts evenly into one another and does not trap air bubbles. Flatten out the base and then spread the hardener over it with a flowing action. Continue this spreading action from alternate sides and working towards the centre until an even colour shows that a good mix has been achieved.

Preparation of the panel surface is important, too. Sand back to sound paintwork to provide the usual feather edge. Sometimes, as with galvanised substrate, it is necessary to sand down to a ring of bare metal around the area to be filled to ensure good adhesion of the filler.

Thorough cleaning is essential to remove the sanding dust. If dry sanding is being used the surface will also be contaminated with stearate powder, which is used to lubricate sanding discs. Draw off the bulk of the dust, wipe over with a tack rag and finish with a suitable de-greaser. To avoid corrosion, do not touch the surface with bare fingers after this treatment. If the panel is likely to be cooler than the ambient temperature (the temperature of the surrounding air) it must be warmed to prevent condensation forming. A hand-held infra-red heater is ideal for this job. Filling must be done while the surface is still clean and dry, and certainly within the hour. Allow ample time for curing or use an infra-red heater if this is permissible.

Where aluminium bodies are concerned new products and techniques have been developed and this will, no doubt, continue. Use products approved by the vehicle maker and follow the repair instructions, especially if body warranties exist. Cleanliness is of particular concern. Within a short time of preparing bare aluminium, the metal develops a layer of oxide resulting from contact with the air. This invisible layer must be removed if problems with adhesion are to be avoided.

Here is your checklist of points to observe when carrying out any filling:

- Prepare the area to be filled in the specified manner.
- Use the filler appropriate to the job and approved for it.
- Use filler and hardener that has been properly stored in sealed containers.
- Mix the parts together on a clean palette using a clean, flexible spatula. Both should be of non-absorbent material.
- The ratio of hardener to base filler must be measured as accurately as possible.
- Fold the hardener into the filler carefully, aiming for a thorough mix without air bubbles.
- Brushed or sprayed fillers are also measured and mixed cleanly and thoroughly.
- Clean the surface to be filled by de-dusting and chemical cleaning.
- Warm if necessary.
- Apply immediately in several layers, taking care to avoid trapping air.
- Shape to slightly proud.
- Cure as recommended.

After curing, the filler can be sanded to shape. Polyester fillers absorb moisture so they must be dry sanded. To obtain a high-quality finish, sanding must be done in several stages to progressively reduce the depth of surface scratching. First, spray on some matt black guidecoat. Using P80 or P120 abrasive, sand off the high-spots with a random orbital sander. De-dust and chemically clean the filler and adjacent area.

Apply the fine filler, in several coats if necessary, and allow it to cure. Spray on guidecoat and sand to shape, using P180 followed by P240 sanding discs. The sander is chosen to suit the curvature of the surface. On the sharpest curves it might be best to finish with a hand sanding block, fitted with P240 abrasive.

The vehicle is now ready for the priming process, which must be completed before the sealing can be applied.

Sealing

Sealing is an important part of anti-corrosion treatment, particularly on vehicles with active warranties. In some spe-cialist applications a sealer is combined with the role of an adhesive, as it is for bonded windscreens. In this section we shall be concerned with those products that are added after assembly purely to prevent the ingress of moisture. Satisfactory protection will only be obtained by using the correct product and applying it competently. With vehicles under warranty, sealants supplied or approved by the vehicle maker should be used to avoid problems.

The wide range of sealants available may be considered in two main groups; non-paintable and paintable. Both types of sealers are available for machine or hand application.

Sealers which cannot be painted should only be used in those areas where appearance is of no importance and where the user of the vehicle is not likely to touch them. They are mainly intended for bonding and sealing components such as folded flanges, like those around door shells.

Where the sealing can be seen or touched, such as around the outer edge of door skins and inside luggage compartments, the sealer must be of a paintable type. These invariably set a skin and do not react with normal paints.

The appearance of the sealing in these visible places is also of some concern and an OE or factory finish (OE means Original Equipment) is easier to achieve with machine delivered compounds. A guide arm, which can be made from a tubing offcut, is almost essential to obtain the smooth and even appearance of the original. But there is always room for individuality. When the writer was visiting a car plant, the peculiar shape at the end of the door sealing was found to be made by the factory operator smoothing off the end with his thumb!

A good result can be obtained with manual applicators provided that the same sort of guide is attached to the applicator itself and a steady speed is maintained in conjunction with even finger pressure on the trigger.

Many of the factors that affect the application of fillers also apply to sealing:

- Cleanliness is just as important if the material is to adhere properly.
- The temperature of the panel may also result in moisture from condensation and will need drying off before sealing can be done.
- Warming the sealer will help to ease the application if the cartridge is cold.
- Cut the nozzle to the best opening size for the job.
- Never use sealers beyond their 'use-by' date.
- You may also find it helpful to set up the job to allow ease of movement and control. Putting the vehicle on a hoist, or a door on a support frame, may make all the difference.

Fitting bonded glass

Flexible leather gloves are essential for any work with glass, together with normal protective clothing. It is advisable to wear some form of eye protection to guard against a small particle entering the eye. Some of the bonding materials contain a very small quantity of isocyanate. In a normally ventilated bodyshop this is unlikely to cause any problems except to those who have become sensitised to them. These personnel should use a suitable breathing mask as a precaution.

Glass that is bonded to the bodywork is designed to add to the overall strength of the bodyshell. Glass is not very flexible but it does have tremendous strength in compression. An ordinary jam jar, for example, can support the weight of a man quite easily. To provide that degree of support to the vehicle body demands bonding of the highest integrity. Any weakness may well reduce the 'roll-over' protection built into the bodyshell.

Successful bonding that will remain effective demands that mixing, where necessary, and applying the adhesive sealer, as well as mounting the glass, are carried out correctly. There are three main conditions that must be met before any attempt is made to fit the glass.

First, obtain the correct adhesive sealer. If the body is under warranty use a material that is supplied by the vehicle maker or that is approved by them (see Figs 53 and 54). This is particularly important where the body is made of aluminium. It may be necessary to use an adhesive sealer which has low electrical properties to avoid contact corrosion. Next, check if there is a 'use-by' date and if it may still be used. Some of the bonding materials age, even in unopened and sealed containers. They will not cure properly if used beyond the time limit so never use out-of-date bonding. Then check the preparation instructions, especially if the material is new to you.

The other conditions relate to preparation. It is usual to cut back the previous bonding to no more than 1 mm in thickness. In doing this the paint film may have been penetrated in places. In the case of a new bodyshell there may be some handling damage around the aperture. Whatever the reason, any paint damage must be rectified to allow bonding to take place at those points and avoid the early onset of corrosion. Where the vehicle is still under a body warranty, the maker will probably ask for a particular treatment to be used. Invariably, all such treatments are completed with topcoat. Special glass bonding primer must then be applied to these areas.

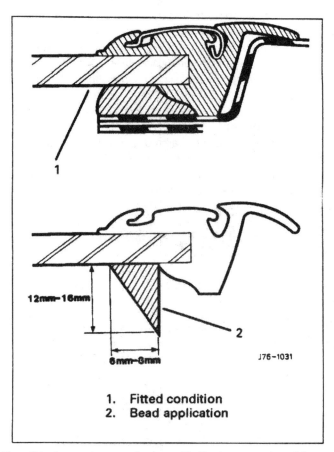

1. **Fitted condition**
2. **Bead application**

Figure 54 Automatic screen depth provided by the surround moulding on the Jaguar; note that the bead is a slightly different size and shape to the Saab

The glass, too, may need preparation. New glass often needs priming to ensure a good bond. Re-used glass may also need repriming. Although the aim once again is to leave a thin layer of old bonding, priming will be needed where it has been removed back to the glass. Cleanliness is vital in all these operations, and it is particularly important that the trimmed surface of the old bonding be kept perfectly clean. Contamination of any kind may weaken the bonding or even prevent it altogether.

When the initial preparations are complete check the instructions carefully, especially if you are using a fast setting adhesive sealer for the first time. Manual mixing of two pack components must be very thorough and at the same time it may have to be done quickly. An automatic mixing applicator takes care of the mixing for you. Curing can start in as little as fifteen minutes after the two parts have been mixed so fast installation is necessary. Make it a practice to do all the preparation beforehand. All the available time can then be used to apply the bead of adhesive sealer and settle the glass into position.

One of the latest 'high position tack' materials is faster still. It is warmed before application and applied immediately. Such adhesives are similar to those used in production where, as the car is moving along the production line, there must be almost instantaneous bonding. The latest products give the benefit of 'drive away' times as short as fifteen minutes. The disadvantage is that there is virtually no time for repositioning; the glass will bond almost as soon as it is lowered into position.

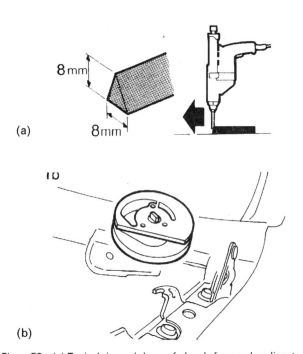

Figure 53 (a) Typical size and shape of a bead of screen bonding, together with (b) the screen adjusting cam on the Saab 900

Whatever type of adhesive sealer is used the bead is laid on the glass. An even bead is more easily obtained using a pneumatic applicator. A firm support for the glass is essential and a purpose made jig is best. Whatever is used to support the glass, prevent scratching by ensuring that there is no grit on the contact point surfaces.

The adhesive sealer must be able to stand up in a peak without any sagging. This shape is created by a cut-out in the cartridge or applicator nozzle. A guide, which may be an extension of the nozzle or a separate attachment, is needed for a smooth and evenly placed bead. This is particularly valuable when using adhesive sealers with limited pre-curing time.

Any positioning spacers must be checked and adjusted. The glass is carefully lifted into place and settled into its final position. If the process calls for the use of clamps, fit these into place and tension them as required before curing commences. Sometimes the glass must be pressed into a particular position to provide the correct aerodynamic shape for smooth airflow. This position is not always achieved automatically, so always check in the repair instructions for the vehicle. A pre-prepared 'go/no-go gauge' is the easiest way to achieve the correct position.

What happens next may well spoil an otherwise perfect job. The glass and the surrounding structure will only become fully integrated at the end of the curing process. With some adhesive sealers this may be as long as five days, depending upon temperature and humidity. There is a stage before this where it is possible to use the vehicle. Before this time it is essential that the vehicle is not subjected to body flexing stress or the bond will be weakened or even destroyed. Ideally, do not move the vehicle during the initial curing period, which may be as long as twenty-four hours. Resist the temptation to clean the inside of the glass during this time. Outward pressure from inside is disastrous.

There are variations on the theme of bonding. One method of fitting the rear screen to a cabriolet employs a built-in heating wire to melt the pre-coated adhesive for installation and soften it for removal. In all special applications the vehicle maker's instructions must be carefully followed, including the use of any special tooling.

Fitting mechanically mounted glass

The traditional way of mounting glazing by the use of 'H' section rubber channel is falling into disuse with the requirement for lighter and stronger bodies. The cord method of installation has been in use for many years. This is the procedure:

- The moulding is placed around the glass.
- A cord is inserted into the bodywork recess of the 'H' moulding and wrapped around to overlap on a long edge, in the case of windscreens the lower side.
- An assistant lowers the glass into place and the ends of the cord, one at a time are carefully pulled inwards. This will pull one side of the rubber over the flange.
- The glass must be held close to the aperture and coaxed into place from the outside.

- The cords must be pulled carefully and steadily to provide a controlled movement of the rubber and to avoid damage.

There are some important points to watch if problems are to be avoided:

- The size of the glass relative to the aperture is important. Bodies always flex in use and there have been many instances of glass breakage due to insufficient space between the glass and the body aperture. Always make sure that there is a clearance equal to the thickness of the centre rib of the rubber surround, between the edge of the glass and the opening. Remnants of old glass remaining in the rubber surround have the same effect.
- Always check that a windscreen surround rubber which is to be re-used is completely clear of particles and in good condition.
- Any damage to paintwork or corrosion that becomes visible after removal should be dealt with. Where there is an anti-corrosion warranty in force this must be carried out to the standards of the vehicle maker to avoid problems in the future.

Some bonded glass is mechanically fitted and retained in the vehicle. The glass is bonded to a frame which is held in place by screwed fasteners or rivets. Where multiple studs, bolts or screws are used these will be of small diameter and must be treated carefully. Prepare the mounting and the glass frame carefully. When in position fit all the retainers and take up the slack evenly so that there is identical tension on each of them. Then start to tighten them one by one a little at a time, moving to the opposite fixing on the other side each time. Do not over-tighten: if there is a torque setting – use it! Some recent aluminium bodies use bolts and rivets to retain some glazing. Be careful to use properly treated screwed fasteners and rivets where this is a requirement and follow assembly instructions.

The door glasses are invariably mounted in channels to allow for opening and closing, with the lifting device attached to the bottom edge. Vehicles without an upper door frame usually have adjusters to allow for accurate setting of the window to the door seals. Unless full set-up instructions are to hand, reassemble to the marks made on dismantling and check carefully where settings need to be altered. Adjust each one, a little at a time. Where the door glass seals against the door aperture seal, the glass position is usually set last.

Fitting electrical components

This may range from replacing a lamp to a complete installation during a bodyshell change. In either instance, cleanliness and painstaking care are the secrets of success.

The first and very important part of the electrical circuit to check are the earth points. The body is used, with rare exceptions, as the part of the circuit that returns current to the battery. It is known as the earth return system. Poor earth connections have peculiar results on most vehicles. Lamps flash, come on when they are switched off or are just very dim.

Poor earth connections after body repair or repaint occur because the screwed connections or spade terminals have been painted over or are dirty. The mating surfaces of all earth connections at the point of contact must be clean for current to flow. The same is true of connectors. Storing the wiring harness carefully should have taken care of this aspect for you but check connectors, before pushing them home, to make sure.

One other aspect of earthing concerns radio aerials. They must be earthed effectively to provide good reception. Scratching away protective coatings and paint, particularly under front wings, is counter to the anti-corrosion measures provided on many vehicles. One way of overcoming this possible cause of corrosion is to attach a separate earth lead from the aerial mounting to an adjacent earth point on the vehicle.

Although checking for damaged wiring is part of dismantling, always double check as you install to guard against oversights. Cable clips or anchorage points should be used whenever these are provided to minimise the chance of noises. More importantly, wherever cables pass through panels they must be properly protected by grommets that are in good condition. Insulation failures at such places are a major cause of vehicle fires. In the engine compartment, loops of harness or individual cables that can touch hot metal must be properly supported by strapping or clipping to suitable fixtures. Watch too for signs of wear and tear at heavily worked places. One example is the heated rear screen cables where they pass into the tailgate on some hatchbacks.

The installation of 'black boxes', the miniature electronic controllers, should be done most carefully. The multipin connectors can normally only be fitted into their correct socket and the right way round. Make it a rule to check as you assemble, certainly before attempting to switch circuits on. Look too at the terminals and pins inside the connectors and on the black box. They can be easily bent out of position or may have been pushed back into the holder.

Consumers such as lamps, heated screens and electric mirrors will all have provision for an earth return. Check these carefully and make sure that all grommets and rubber boots are sound and in place. Some lamps, usually rear lamp assemblies, are mounted on a seal or with sealing cord. To prevent water ingress as intended it must be in sound condition and sealing all round the cluster.

Headlamp assemblies are mounted directly and accurately onto the front panel so that the full range of headlamp adjustment is available (see Fig. 55). This is particularly important on those vehicles with headlamp range adjusters.

Sunroofs are another mechanism which may be electrically operated. The earlier comments on connection and earth points apply to them too. Be careful about checking the type of sliding roof that drops into a recess in the roof panel, either electrically or manually operated. Opening a sliding roof that has been incorrectly adjusted can damage the paintwork. Obtain a copy of the adjustment instructions and set it up in the correct sequence.

Air conditioner systems must be reassembled carefully to avoid leaks and ensure correct operation. Air must be evacuated and the system charged with the appropriate refrigerant and oil. Observe the safety precautions outlined at the

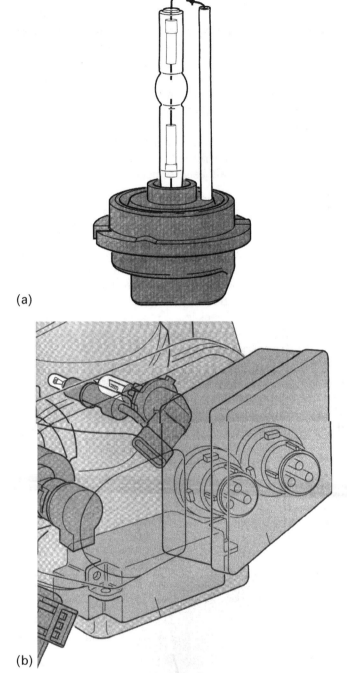

(a)

(b)

Figure 55 (a) the gaseous discharge headlamp 'bulb', together with (b) the starter and control units on the Audi A8; gaseous discharge 'bulbs' are expected to last for the life of the vehicle
High tension current is generated to start the lamp and the usual precautions against electrical shock must be observed

beginning of this chapter. The two most important points to remember about R12 are:

- This gas is heavier than air and will flow to the lowest point. If that is a pit in which someone is working, they may well be asphyxiated.
- If the gas is drawn through a flame, as with someone smoking, deadly mustard gas is produced.

Whatever the electrical component that has to be fitted the electrical conditions are the same for all. Feed, control and earth connections must always be clean and tight.

Fitting external trim

Most external trim is made of one of the many types of plastic. Attachment is by mechanical means, mainly clips, or an adhesive or adhesive tape. There are some special fittings such as aerodynamic designs of roof guttering or water channels designed to reduce drag. These too may be secured by clips or sometimes by such devices as 'welded-on' studs (see Fig. 56).

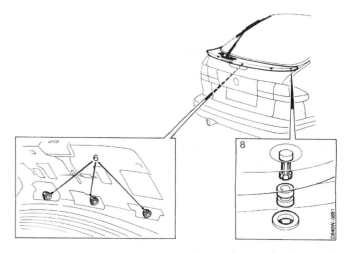

Figure 57   The rear spoiler mountings on the Saab 900; the numbers refer to the text in the manual

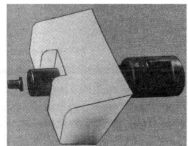

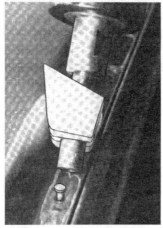

Figure 56   Some vehicles have welded-on studs to retain external trim; this sequence shows the special tools for aligning and welding the pins that retain the roof mouldings on the Volkswagen Golf

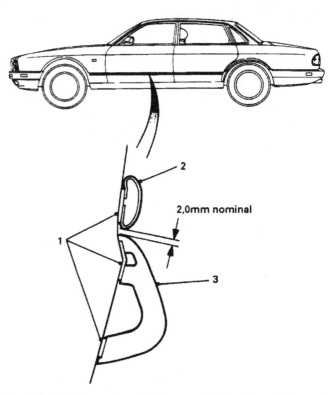

Figure 58   This diagram from a Jaguar manual shows side mouldings and the mounting positions for various models
1   Adhesive tape
2   Moulding, Daimler/VDP
3   Moulding, all models

There are two main areas of concern. The trim must look right and it must do the job. Bumpers should be carefully positioned to be central and be equally spaced from nearby panels (see Fig. 57). Side trims should align along their length and be carefully placed in relation to swage lines (see Fig. 58). Water channels and door seals are fitted first and foremost to do a job. But keep an eye on appearance as well.

Trim retained by clips can be the cause of corrosion. Expanding plastic clips pushed into holes in side panels can chip paint from the edge. Such damage will almost certainly cause corrosion which can gain a foothold and flourish while it is hidden behind the trim. By the time it is seen the opening in the panel is likely to be so badly damaged that a welding repair is needed. This may be avoided to some extent by making sure that the paint is fully cured before fitting up, warming the plastic or nylon clips, giving them a smear of preservative wax or grease before fitting and careful alignment of the clips against the mounting holes.

Mounting by adhesive or adhesive tape calls for cleanliness in preparation and care in alignment. Check the trim for fit against the panel and warm it to set the shape. When using hot air blowers, always ensure that the jet is kept moving to avoid hotspots.

Fitting internal trim

Assembling interior trim is, for the most part, a matter of clipping and screwing the various components into place (see Fig. 59). Provided that damaged trim and fastenings were noted at the time of dismantling and have subsequently been obtained, the work should be a straightforward job of reassembly in the reverse order (see Fig. 60).

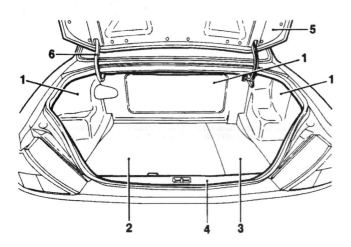

Figure 59 If the door sill hurts your back this may be the answer!

Figure 60 The trim of luggage compartments varies greatly; here is the detail on the Jaguar XJ Series
1 Trunk liners
2 Removable floor board
3 Battery cover
4 Treadplate
5 Trunk lid
6 Hinge

Most present day cars are fitted with stiff headlinings. Older models and cabriolets or convertibles have soft linings. There is nothing particularly difficult about dealing with these if detailed instructions are available. You will need the correct tools and materials, of course.

Most vehicles have sound-absorbing panels fitted to the inside of panels which may resonate. They are also available for fitting after purchase. As new they are invariably self-adhesive. Even if they have been removed successfully it does not follow that the adhesive is now capable of holding them in place. If, in all other respects, they can be refitted a suitable impact adhesive will normally be effective. Other noise damping materials are in use such as liquid coatings for brush or spray application (see Fig. 61).

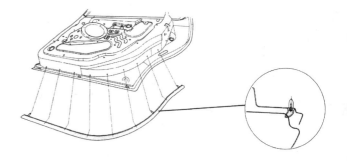

Figure 61 The additional bottom door seals on the Ford Mondeo dampen noise and provide water splash protection

Protective padding to minimise the consequences of side impacts is also being used in some models. It must be refitted or renewed as appropriate.

Many doors are fitted with waterproof membranes (see Fig. 62). Their purpose is to guide water dropping down from the window above to drain away through openings in the bottom of the doors. They are an extremely important part of anti-corrosion protection and must be renewed if they are torn or are too fragile for further service. Replacement membranes are usually fixed in place with double-sided adhesive tape.

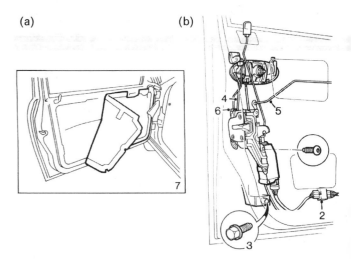

Figure 62 (a) The door components and (b) water barrier of the Saab 900; water barriers are also used on other makes and must always be kept in sound condition

Pyrotechnic devices, such as airbags and explosive seatbelt tensioners, should be fitted as late in the assembly process as possible. When these are being installed the battery must be disconnected to prevent personal injury. When activated in an accident, these devices normally counteract against opposite forces. Activation when someone is crouched or stooping near them could result in injury. When the battery is reconnected for the first time on cars with airbags or explosive seatbelt activators, make sure that there is nobody inside and the doors are closed.

Here are some useful tips that may be of help in trim fitting work:

- The fast thread, self-tapping screws often used in this work can be started in their old thread by backing them until a click is heard or felt. This is the thread start engaging. The screw will then drive in very easily.
- Where there are screws of varying length, be sure to put short screws in the position from which they came. A long screw may damage another component or snag wiring.
- Start all the screws in their holes before tightening. Then drive them fully home while holding the component in the correct position.
- When using a power screwdriver be careful not to over-tighten.
- When assembling a complex build-up of mouldings and panels, such as a dash assembly and closing panels, it is often worth doing a 'dummy run'. This will reveal any snags or alignment difficulties.
- It is very difficult to know precisely where foam padding was installed if you did not strip the vehicle. As you build-up, check if cable harnesses are likely to touch other parts and so vibrate. Clip them away or pad them with foam. Be particularly careful when inserting radios and glove boxes.
- The battery must be disconnected when installing air-bags and pyrotechnic seatbelt tensioners to avoid accidents.
- Make sure that nobody is inside a car with airbags when connecting the battery for the first time.

Chapter 4

Repairing Body Panels

Preparing in situ Panels for Repair

Health and safety at work and PPE

The repair of body panels involves hazards that have been discussed already, such as the edges of cut metal components and the risk of burns from sources of heat. The other major health risk that arises from panel preparation is that of dust. There may also be fumes from the repair of plastic components. Here is your reminder checklist of protection, for yourself and others (see following page).

Dry sanding is now the accepted way of flatting finishing materials or to remove them for the purposes of body repair. This method has come into use because it is much faster. The danger arises from the size of the particles that the sander cuts away and the nature of our lungs.

The purpose of a lung is to replenish the oxygen in the blood supply. The tubes of the lung resemble a tree, with a thick trunk, smaller branches and some 250 000 twigs. On the end of the twigs, like leaves on a tree, are 6 million bubble shaped cells called alveoli. Their job is to exchange fresh oxygen for carbon dioxide. The lungs do have a mechanism for collecting some dust and other foreign matter, which is eventually coughed up (see Fig. 63).

What the lungs cannot cope with is the size and quantity of the dust produced in dry sanding. Many of the particles are between 2 and 5 microns in size (a thin human hair is 30 microns in diameter) and it is these tiny, invisible pieces that lodge in the inner recesses of the lungs. Research has shown that up to 20 million of these particles can be inhaled every minute during heavy sanding work without a mask. Continual, unprotected exposure to this hazard will almost certainly result in lung disorders such as fibrosis or pneumoconiosis. The only certain way to overcome this risk is to provide efficient extraction from the tool itself. PPE can then easily cope with any stray particles.

Housekeeping

Good housekeeping contributes to health and safety too. To be skilled at your work is good. To be skilled and fast is better. To be skilled, fast and a clean worker is a sign of a true professional. Keep your workbay clean and tidy. Save time and cut down hassle by following the old saying 'a place for everything and everything in its place'.

Legislation also has something to say about housekeeping in the form of the Waste Disposal Regulations. There is also a regulation that imposes what is called a 'duty of care' on all those who handle, store or dispose of waste. That means you, as well as your bodyshop. You do not need to

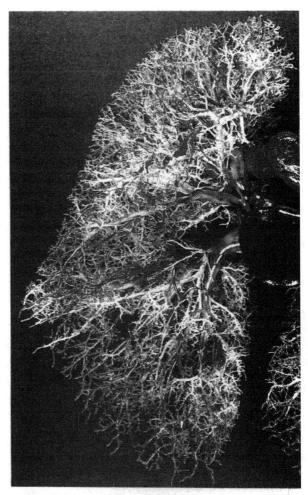

Figure 63   The 'tree like' passages of a human lung

mishandle waste to be prosecuted; not complying with your 'duty of care' is an offence in itself.

Waste from the panel shop that comes under the Waste Disposal Regulations will probably consist of waste paper and cloth wipes, plastic, empty or part used cans and used or partly used containers of hardener. All this material must be stored in closed containers until the licensed carrier collects it. Liquid waste, such as solvents, must be poured into enclosed containers. There are two special concerns. Materials such as the organic peroxides found in catalysts must be stored separately from other waste. Rags or wipes with organic solvents from thinners on them should be kept damp with water to avoid self-ignition.

The regulations concerning the disposal of paint stripper residues are particularly strict. Great care must be taken in the workshop and special arrangements may need to be made regarding disposal by a carrier.

Hazard	Protection
The cutting and drilling of sheet steel may cause injury	Use eye protection against flying particles to BS 2092 Grade 2 or EN 166 grade 'F' Riggers' gloves Industrial overalls
Pressurised gases such as compressed air can be extremely dangerous if used in an uncontrolled manner	Compressed air and gases must only be used with equipment that provides full shut-off and directional control Normal workwear provides no protection against pressurised gases
Mains electricity is as dangerous as pressurised gases if used carelessly or the equipment is poorly maintained	Plugs, cables and equipment must be in good order without defects Cables and plugs must be protected from cuts, chaffing, crushing or any form of tension
Dry sanding produces millions of minute particles which can lodge in the inner recesses of the lungs	Use dust masks to BS 6016 (disposable), BS 2091 (rubber facepiece) or EN 149 class FFP 2S; see text for further information
Dust and particles escaping from shot blasting operations	Overalls, gloves and full hood with eye protection to BS 2092 Grade 2 or EN 166 grade 'F'
Fumes produced during heating operations or plastic welding	Cartridge respirator to BS 2091 or EN 149 appropriate to the fume involved Leather gloves
Plastic, material or foam inadvertently burnt during welding or cutting operations may give off toxic vapours	Every care should be taken to avoid the burning of plastic, material and foam Good ventilation can cope with non-toxic gases released through accidental ignition Large amounts or toxic gases call for breathing apparatus
Other people in the workplace will also be affected by dust and fumes The accumulation of dust also has an adverse effect on the quality of paintwork	Dust and fume extraction, preferably by a centralised system, to reduce levels well below Occupational Exposure Limits General extraction in no way replaces PPE
Excessive noise exposure levels, either peak or on a continuous basis	This is the most difficult danger to assess. Employers must check noise levels and provide protection where they exceed limits, either briefly or continuously Workers must, by law, wear ear protection when it is provided
Treating rust with an acid based metal cleaner	Full acid protection must be worn including gloves, boots and a face visor A respirator with a suitable cartridge may also be needed
Flying particles from wire brushing, panel beating and similar work	Goggles or a visor to BS 2092 Grade 2 or EN 166 grade 'F'
Flying hammer heads	Examine hammers regularly Renew or refit handle with correct wedge as necessary

Sources of information

Trade journals often carry articles or whole supplements on elements of the work in a bodyshop. These can be invaluable, especially in keeping you up-to-date with tools, equipment, materials and techniques.

Vehicle protection

Protecting your customers property and the other vehicles that are usually nearby is essential if damage is to be avoided. It can also greatly decrease the time taken to valet the vehicles on completion of repairs.

The outside of the vehicle you are working on should be covered except for the area of the repair. Use fireproof blankets where appropriate. Quite apart from reducing general dirt, it is important to prevent hot metal particles from landing on the paintwork, glass and the interior trim. Even cold metal particles can gather into crevices and corrode there in the course of time. The same protection should be given to the other jobs nearby.

Check the inside of panels which are to be heated to see if there are electrical components, noise deadening panels,

Job	Source of information
Sanding	Abrasive makers' information Paint makers' information Training information
Choosing equipment for sanding	Equipment makers' information Abrasive makers' information Training material
Selecting dust masks, goggles and ear protectors	PPE makers' information Health and Safety Executive information Specialist suppliers' catalogues and training material MIRRC (Thatcham) information
Panel beating, shrinking and dent pulling	General panelwork information Equipment makers' information Vehicle makers' manuals or training material (special body requirements)
Recognition and repair of plastics	Vehicle makers' information MIRRC information and training material Equipment makers' information Materials suppliers' information

injected foam filling or other items which could be damaged or ignite. Remove, or move, those which can be released. Injected foams or the remains of fused deadener panels may pose risks which call for special safety precautions.

Do not overlook wheels and tyres. Vehicles can often be seen undergoing repair with wheels and tyres exposed to the debris being removed from the bodywork. If wheels cannot be removed and stored, they should always be protected.

Abrasives standards and grading

Abrasive papers are graded by the texture of their surface (see Fig. 64). Low numbers, such as P24, indicate a coarse grit which can remove paint very quickly and cause deep scoring. As the grit number progressively rises, the cutting action reduces and the finish improves. The finest grade is P2000. There are three main abrasive standards in use in the developed world. North and South America use the ANSI standard, which is completely different to the others. The JIS standard applies in Japan and Korea and this is very similar to the European FEPA standard (Federation of European Producers of Abrasives). In our FEPA standard the grit number is pre-fixed with the letter 'P'. The FEPA grading system is also used in Australia.

Despite having a standard, caution must be exercised whenever a different supplier is used. Even a different type of abrasive from the same supplier can vary slightly. The advice from the industry is to stick with those with which you are happy. If you must use a different make or a different method of fixing from the same maker, test out the results on a scrap panel before attempting the job.

Sanding should always be done in stages by changing up a grade or two at a time. In this way a coarse grade can be

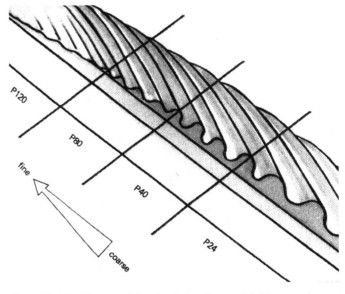

Figure 64 The 'P' range of abrasive designations and finish removal

used first to break into the surface. As the material is reduced in thickness the aim should be to contain the depth of grooving within the material thickness until a smooth finish is achieved just before or at bare metal stage. This prevents damage to the metal surface and provides a clean and gradual reduction of the paint thickness (feather-edging) of the existing paint. It follows that if a spot repair is being done cutting should begin well within the outer limits of the repair area. As each progressively finer grade is used the area is enlarged to provide a gradual change in paint thickness.

Grades used during dry sanding preparation normally range from P80 through to P320. For the preparation of colour coats a grade as high as P800 may be used. When rectifying defects in topcoats abrasives are used wet and range between P1500 and P2000.

Wet flatting follows the same principle although a finer grade of paper is used for similar jobs. Typically, a P600 paper might be used in wet flatting instead of a P320 when dry flatting. While the work is in progress the water carries away the residues so there is little risk of breathing in sanding particles. However, care must be taken when cleaning up the dry dust from the floor. This problem does not arise if a wet deck, a grid floor with drainage, is used.

The other material used for fine surface preparation is the 'woven' abrasive pad universally referred to as Scotch-brite℗. This is used for flatting existing coatings to provide a keyed surface to which fillers and subsequent paint coatings can adhere.

Equipment

Sanding by hand, whether wet or dry, is best done on large areas using rubbing blocks whenever possible (see Fig. 65). These are available in a range of shapes and sizes. In the interests of health and safety when dry sanding, use dust extracted files and blocks whenever you can (see Fig. 66). Extracted blocks will need abrasives with the correct extraction holes punched into them. Dry, machine sanding is a much faster and certainly the least laborious way of flatting. Machines connected to an efficient, centralised dust extraction system (see Figs 67 and 68) are quite safe to use. None the less, a suitable dust mask should always be worn as a precaution (see Fig. 69). On large, flat panels it is best to use a half sheet, flat bed orbital sander to cover the area quickly. Work on car panels mostly involves slightly curved surfaces and the D/A (dual action) random orbital sander should be used. The combination of rotation, an eccentric path and a random orbit provides for even sanding with minimum heat build-up. The orbit of these machines should not exceed 6mm. For preference they should also have internal extraction and be connected to a centralised system. As with hand sanders, the abrasives used should be punched with extraction holes in the pattern to suit the machine. Never be tempted to use machine discs by hand for final finishing. A cut of 3.6 microns by machine may easily become 7 microns by hand.

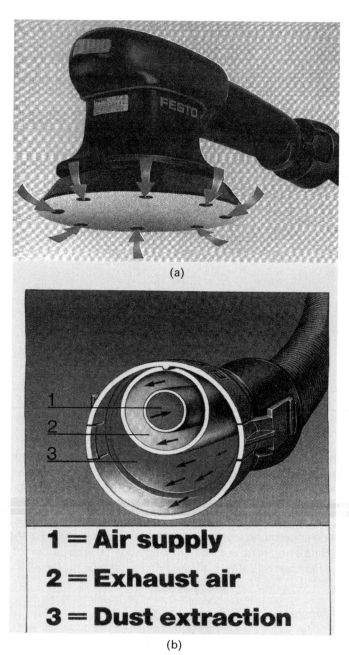

(a)

1 = Air supply

2 = Exhaust air

3 = Dust extraction

(b)

Figure 66   (a) Atmospheric pressure filling the depression inside the sander carries away the dust of sanding; (b) in this particular unit the connecting hose provides the air power and carries away both exhausted air and the dust of sanding

Rust must be removed completely if it is not to reappear. One way of doing this is by shot blasting. Small, hard particles are fired by compressed air at the rusty surface. This breaks off the rust scale. Mobile shot blasting units are available driven by compressed air. Some form of screening, with a method of collecting the used shot and waste, will also be needed.

Removing finish, sealers and corrosion

Abrasives or chemicals may be used to remove finishes for the purpose of rectifying damage or corrosion. Most bodyworkers find the 'no-nonsense' dry abrasives approach the best for all normal jobs. It is fast and relatively clean.

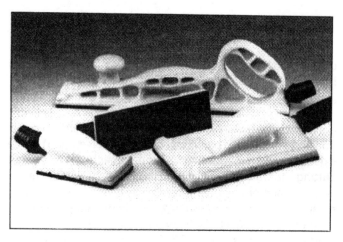

Figure 65   Hand flatting is more easily controlled using properly shaped, dust extracted blocks and files

Figure 67 The power plant of a centralised dust extraction facility. The dust is gathered in the plastic sack

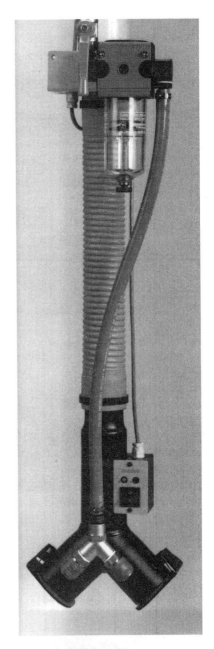

Use a half sheet, flat bed sander for large, flat areas and a random orbital sander for normal car work. A P80 grade abrasive will remove the bulk of the paint, finishing with a P120. Where a local repair is involved, space must be left for feather-edging with finer grades.

Here are some important reminders about sanding down:

- When about to dry sand ensure that the extraction system is working correctly.
- Have enough suitable dust masks handy to see you through the session.
- Wear gloves.
- Use new discs to break the surface. Renew discs as soon as the cutting action falls away.
- Ensure that extraction holes line-up correctly.
- Place the sander in position before switching it on.
- Keep the speed down.
- Let the weight of the machine provide pressure on horizontal surfaces; use similar pressure for vertical panels.
- Keep it moving to avoid heat build-up and work progressively over the area.
- Stay inside the final repair area until you are ready to feather the outside edges.
- Work away from sharp edges or curves unless stripping these areas.
- Every effort should be made to preserve surface galvanising and maintain the panel thickness (see Fig. 70). Never use a grade coarser than P180.
- Change up to finer abrasives in stages no more than two grades apart.
- When using wet flatting for preparation use clean water to avoid contamination.
- Change the water for every job.
- Rinse and leather off the residue of wet flatting before it dries.

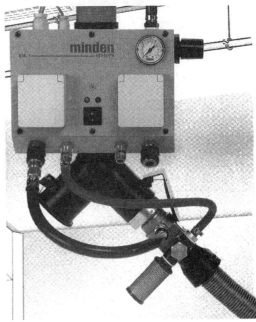

Figure 68 Pendant extraction and supply modules

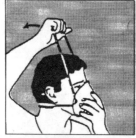

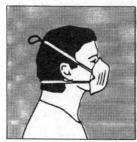

Figure 69 How to fit a face mask correctly; in picture 4 the mask is being pressed into the recesses on both sides of the nose

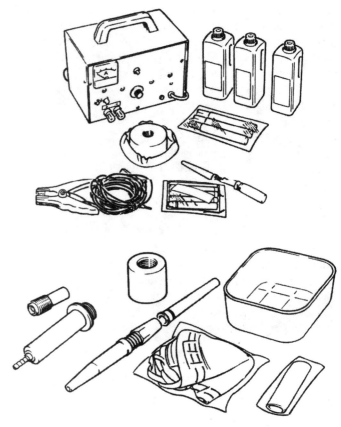

Figure 70 The Peugeot re-galvanising kit for reinstating corrosion protection after repair

Rust may be removed if it is only on the surface and penetration or significant weakening has not occurred. Chemical removal may be carried out using a metal cleaner based on phosphoric acid. Although harmless to steel, aluminium and most cured paintwork, acids will attack zinc coatings and should not be used on galvanised steel. Metal work treated in this way must be thoroughly washed before the acid has dried. When using these materials, full acid-proof PPE, including suitable footwear, must be worn. Follow the treatment instructions carefully.

Shot blasting is the alternative method for dealing with rust. It is imperative to remove all the rust to effect a cure: the tiniest trace will reactivate under the paintwork and eventually show through. The spent shot and residues from this work must be contained without any possibility of injuring people or damaging cars. A full hood and visor will be needed together with overalls and suitable footwear.

Surface treatments

There are other treatments given to repair surfaces. Here we are going to look at some that should be used before or during repair. The first concern is the road dirt and contamination that may be on the vehicle before it enters the repair shop. Second is the treatment of the panel during repair and the measures that will prevent surface corrosion leading to failure of the paintwork.

When vehicles arrive at the bodyshop they will rarely, if ever, be clean. They will carry the dirt of normal use, and substances that have been applied to the paintwork,

Sealers will need to be cut away to carry out some repairs. Where there is a substantial amount, as is often the case around door shells, the bulk can be cut away. The oscillating chisel blade of a mechanical cutter as used on windscreens is ideal. The remainder can be removed with a rotary wire brush or a mule skinner. This should be of stainless steel for bodywork constructed of aluminium. Sealer may often be softened with a hot air blower. Be careful not to overheat any one place. PPE must include goggles or a visor to BS 2092 Grade 2 or EN 166 grade 'F'.

including waxes and polishes containing the worst nightmare of painters, silicon. Contamination with silicon is a disaster for the paintshop. It can be the cause of costly extra work and prove very difficult to eradicate. Prevention is always better than cure. Every vehicle should be thoroughly cleaned before entering the bodyshop, preferably with a pressure washer and cleanser. The areas to be repaired should also be chemically cleaned with a solvent de-greaser to ensure that any remaining contamination is not carried onto the repair surfaces. Chemical cleaning is repeated throughout the repair and refinish process.

During the flatting and filling processes the metal surface, though visibly clean, may easily be contaminated. Just one clean finger print is all that it takes. At every process, then, it is vitally important that a chemical de-greaser is used to remove any trace of contamination from the metal and paintwork. Always use lint free, clean wipes. Aluminium too can give problems as it rapidly oxidises on exposure to the air. This oxidation is invisible. If the surface has been exposed and unworked for an hour or more, it must also be cleaned thoroughly with a stainless steel wirebrush.

Reinstating Metal and Plastic Components

Health and Safety at Work and PPE

The principal hazards in this section are those related to using panel beating tools, the heat from shrinking and the fumes from plastic welding. These are covered in the table on p. 46 at the beginning of this chapter. Hydraulic pushing and pulling may also give rise to danger from the escape of hydraulic pressure or the accidental movement of equipment under tension.

Equipment and tooling

Re-shaping dented panels in the traditional way calls for the use of hammers, dollies and spoons. Work on aluminium is best carried out using similar but softer tools made from plastic, wood or aluminium. All beating tools must be maintained in spotless condition on their striking surfaces. Beating may be carried out on all of the smoothed surfaces of dollies. Any slight marking may be removed with emery paper lubricated with oil.

The modern alternative to beating for small dents is the use of the pulling, pushing and heat shrinking processes available with some welding equipment. For example, the single sided spot and rolling seam welder of one make of equipment can be fitted with an inertia hammer that is combined with a weld-on tip. In this way the tip can be welded to the deformed metal, which can then be pulled out with the hammer. By changing the hammer unit for a copper rod, any resulting deformity can then be pushed back into shape, using the heat generated at the tip to restore the molecular structure of the metal. A carbon pencil replaces the copper tip for final dressing (see Fig. 71).

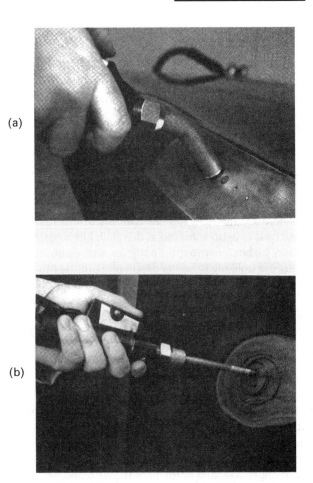

Figure 71 Shrinking dents, using (a) a copper shrink and (b) a carbon pencil

Heat shrinking aluminium is made difficult by the absence of colour change as the metal warms. Because some aluminium begins to change its structure at 200° Celsius, a 150° Celsius thermochrome pencil must be used to avoid overheating the metal. Heat shrinking may also be done with any heat source, of course, provided that it is properly controlled and monitored.

An inertia hammer with a hook can be used to pull out dents by welding-on washers to the damaged area.

There are other hand tools that can help to make your work quicker and easier. Edge setting (joggling) pliers to create the flange for an overlap joint when using a part panel avoid the need for a butt joint. Door skinning is made easier and the finished result looks very like factory production if a flange crimping tool is used. A door skinning rig to hold the door frame firmly is also most useful.

Welding equipment may be needed for a number of these jobs. Pulling out dents with welded-on washers or studs is often a better way of removing dents. There is no need to obtain access to the back of the panel and such is the low level of heat that even anti-corrosion wax treatments will not normally be affected. Welding equipment may also be needed for partial door skin replacements. Some door skins are attached with resistance spot welding and MIG/MAG welding may be needed for joining a part panel.

Body soldering (lead loading) requires a heat source sufficient to heat the body panels. A hot air blower or a propane torch are best. An oxy-acetylene welding torch is less 'user friendly' because of the difficulty of controlling the concentrated heat.

The welding of plastics needs a low level of heat and this can usually be done with a special form of hot air blower and attachments. Not all plastics may be welded and an identification kit may be useful to identify components made from an unknown material.

Reinstating panels and body members by beating

This is not a skill that can be learned from the pages of a book. There are, however, some important points to remember if success is to be achieved in this skill. Flat sheets of metal may be beaten into quite complex shapes but work of this nature is unlikely to be needed in any but the most specialist bodyshops. Panel beating in day-to-day repair work usually involves the elimination of dents, from the inside out or the outside in. To a lesser extent creases in panels may be removed or diminished. Hammer, flipper and dolly are also used for the dressing of flanges during repair preparation.

The tools must be kept in the very best condition if damage to the panel is to be avoided. All working surfaces on hammers and dollies must be smooth and free of the burrs that form around indentations on the faces. Emery cloth and oil are used to restore the polished surface. Panel beaters hammers should not be used for any other work and never on sharp edges. Despite hardening, the surface will soon deteriorate. The security of the handle is important for ease of use and for the safety of all around. Examine the shaft and head union regularly and refit or replace the shaft if necessary. The correct wedge should be used to spread the handle in the head.

Work on softer metals like aluminium demands softer tools. Plastic or nylon are considered suitable. Ideally, they should have replaceable heads as they cannot readily be resurfaced. These are usually of screw-in type. Always check that they are in good condition and secure before use.

Dents are the most common form of damage and occur on all types of vehicles. Many minor dents have no damaging effect on the paint film and its adhesion or on anti-corrosion measures. A flat door panel will readily distort into a gentle hollow with even modest pressure and the flexibility and adhesion of the paint film may be quite unaffected. Nonetheless, they are unsightly and can disfigure an otherwise well kept vehicle.

The force of the impact and the stiffness of the panel determines the nature and extent of the damage. A stiff, curved panel hit with some force is likely to have a deep, sharp indentation. Such damage will almost certainly break or detach the paint film so must be remedied without delay.

In all cases, the metal will have suffered some degree of stretching. Without removing this stretch, the panel will not return to its proper shape or contour.

Beating technique differs depending on the panel material. Dents in steel are treated by beating from the outside of the depression and moving inwards towards the centre. Aluminium should be pressed into position wherever possible but if beating must be done the technique is to start from the centre and work outwards.

It is commonplace to give a light skim with a body file to reveal high and low points. A light dressing in this way will remove only a few microns of metal and will not significantly alter the panel thickness. If beating has not restored a smooth surface, despite a good contour, filler must be used to restore the final shape. Under no circumstances should metal thickness be heavily reduced by filing or sanding in an attempt to remove any indentation. Any remaining finish can be removed with a wire brush or mule skinner.

Reinstating panels and body members by pulling and shrinking

This is the most cost effective method of restoring panels to their original shape. The advantage lies in the fact that access to the reverse of the panel is not needed. Conventional inward dents are drawn out by means of washers temporarily welded to the indentation. The washers are attached by using a special adapter on a resistance spot welding machine. An inertia or slide hammer can then be hooked into each washer in turn and reverse blows used to draw out the dent. The washers are removed and the dent can then be heat treated. This equipment may also be fitted with a reusable, weld-on tip, which enables small dents to be pulled out without any need to change tools. The tip is removed from the panel by a twisting action as each pull is completed.

Conventional forms of heating, such as an oxy-acetylene welding kit or a propane torch, may be used. However, there is a risk of damaging the panel material and a great deal of skill is necessary to avoid excessive heat. The better way is to use carbon or copper shrinking. Carbon shrinking involves passing current from a resistance spot welder through a carbon 'pencil' moving over the surface. The technique is to start at the outer extremities of the dented area and move the pencil quite quickly in a circular spiral towards the centre. Copper shrink is achieved by using a copper electrode with a much larger diameter, convex face. This is primarily used for outward facing dents. Pressure is applied to the peak of the dent so that the combination of pressure and heat begins to draw or pull the metal back to its original shape. As the bump begins to disappear, the copper shrink should be moved in a circular motion, spreading out from the centre. Because the copper shrink is very effective, only the worst of the protrusion is dealt with in this manner. It is usual to change to the carbon pencil for the final shaping.

Reinstating panels and body members by hydraulic reforming

Panels which have been pushed badly out of alignment or have sustained severe creases will need to be brought closer to their normal position before beating or shrinking can be carried out. This is done by using a smaller, lighter form of the hydraulic ram described more fully in Chapter 5 under Body Shell Alignment.

These are available from a number of makers although they are often called Porto-Power®, one of the original trade names. These light, portable units are available with a range of brackets, pushing pads and extensions. These enable the ram to be set-up so that one end is firmly anchored against a solid component, while the other pushes against the deformity in the panel.

These units are remarkably powerful and can move very easily when applied to relatively weak panels. It is important then that a slow and cautious approach is adopted to avoid too much panel movement, too quickly. This is especially true of creases where the initial resistance may need higher pressure. Once that is overcome, pressure must be eased or applied gently to avoid creating further damage. Tension in the metal can be overcome by beating as is described more fully in Chapter 5. Heat has not been used for some years because of its adverse effect on High Strength Steels (HSS). It is of value, however, when dealing with aluminium structures which respond well to heat treatment. To avoid damaging the aluminium by excessive heat, panel temperature must be monitored. The easiest and least expensive way of doing this is with thermochrome crayons or pencils.

Shape assessment

The panel technician is responsible for the contour of the bodywork which has been repaired. During repair work, the eventual condition of the body must be kept in mind. This is particularly important where there are a number of panels joining together. Rear quarter panels often present a challenge where they join with the side members and around rear lamp clusters. The greatest care is necessary to limit the amount of filling that may be needed to provide a smooth shape.

The assessment of the body contours should take place routinely at all stages of the work. It is, quite literally, standing back and viewing the outline of the panel or assembly. A smooth curvature or steady transition from one shape to the next is what is needed. The ends of the fingers trailed across the surface will help in deciding if the shape is smooth. Cotton gloves should be worn to improve the natural sensitivity of the finger tips. Where there are sharp, clearly defined edges they must be crisp and clean. Check the shape by eye from a gradually altering viewpoint. As an example, you could check a front wing shape by standing in front of the bonnet and looking down from above. Then gradually lower your head towards the bonnet. As you do so, the outline of the wing will change. It helps to view shapes against contrasting colours, when the contour will stand out.

Where only one side of a vehicle is being repaired the other will serve as a pattern. Another identical vehicle can also serve as a basis for your assessment. Check small sections of panel by feel or a combination of feel and sight. While keeping the hand flat, lightly trail the fingers across the area to sense the small variations in shape that may be there.

The fit of doors, tailgates and bonnet are, of course, a part of this process. The shape of opening components must be checked against the aperture that they are intended to close. The lines should follow the outline of other body parts, for example the front of a door relative to the 'A' post. However, it must be remembered that on some vehicles the leading edges of such components may be set slightly behind the panel in front to reduce wind noise. It must be possible to obtain such settings (see Fig. 72).

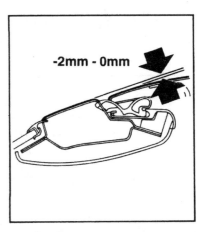

Figure 72   The incorrect positioning of components exposed to the air stream flowing over a car can create wind noise; this setting for the header rail on the Saab Cabriolet is typical

Last but by no means least is the checking of the spaces between the various parts. A repair may be perfect in every other way but if the gap between panels is different on one side of the bootlid to the other the customer will not be happy! The doors seen from the side and the bonnet from the front are the other sensitive views (see Fig. 73).

Door skinning

Some manufacturers make door skins available for those models where a totally new door has a high replacement cost. In such cases it is a relatively straightforward matter to cut all or part of the old skin away and attach a new one.

The door is constructed of an inner frame, to which an outer panel (the door skin) is attached by bonding, spot or MIG/MAG welding, or by a combination of all three. The outer panel has a flange at 90° around the edges. The flange is folded over the edge of the frame to secure the skin. On some early vehicles the skin was spot welded without bonding and only partly folded over. Since the introduction of anti-corrosion measures, door skins are usually bonded, together with some welding. The adhesive serves two purposes. It holds, or helps to hold, the components together and prevents water collecting in the fold by filling the gap. Anti-corrosion measures are completed by sealing the edge of the fold.

The old door skin is removed by grinding through the edge of the fold. If it is also spot welded these too must be drilled out. Additional cutting may be needed depending upon the door construction. Care must be taken to avoid cutting into the door frame itself during each of these operations. The bonded sections are then parted from the frame using a sharp bolster.

The frame is prepared by grinding away any unwanted part of the old panel or welds and then cleaned using a rotary wire brush or mule skinner. Take care to carry out any preassembly preparation, particularly any anti-corrosion measures. These are almost certain to include weld-through zinc primer.

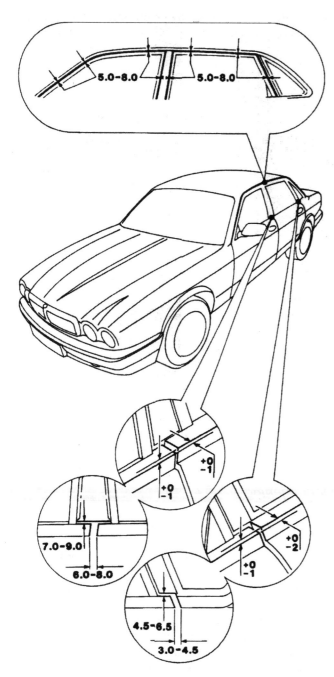

Figure 73 Door setting tolerances on the Jaguar

ing would cause. Some welding equipment is now provided with rollers that enable pulsed resistance welding of seams to be carried out.

Body soldering

Lead loading, another name for body soldering, is a skill that is less in demand and is not even practised at all in many bodyshops. It is, however, a more substantial method of filling with little or no likelihood of problems arising during refinishing or later. It is the only acceptable method where a contour needs to be made good by a substantial filling which might pose adhesion or settlement problems for chemical fillers. It does demand practice on a regular basis for the skill to be fully developed.

Present day 'solders' contain less lead than previously. However, adequate PPE should be used as protection from heat and the hot metal. Waste material must be carefully cleaned up, bagged and put into the correct waste container.

The basic principle is that the solder used must be made soft enough to wipe onto the panel yet not run off. The central skill is the control of the heat in the panel. Too little and the solder will not melt sufficiently to adhere. Too much and the build-up already achieved will run off. Oxy-acetylene may be used but the gentler heat of a butane torch or an industrial hot-air blower is much better (see Fig. 74). The heat source is less concentrated although it is still necessary to take care not to overheat any one part or the area as a whole.

Figure 74 Lead loading or body soldering on the Peugeot using a hot-air blower

Once the preparations are complete the skin may be clamped in position, folded and welded. There are useful hand tools that can make the folding operation easier and give a 'factory finish'. If the skin is adhesive bonded the process will almost certainly be the same as that outlined under Adhesive bonding in Chapter 3. Cleanliness, thorough mixing and sufficient adhesive to fill the gap are the important points to remember. Spot welding, if necessary, should be done before the adhesive has cured. However, the work sequence of folding and welding does vary with make and model so always follow the recommended repair process.

Because of the large, unsupported nature of most door panels, part panel replacement should involve the use of an overlap (joggled) joint where the new panel joins the old. This is usually welded by resistance spot or MIG/MAG plug welding to avoid the distortion that MIG/MAG seam weld-

The aim is to obtain a slightly proud build-up. When it has cooled, the excess solder is filed back until the correct contour is achieved. Under no circumstances should mechanical sanding or fine abrasives be used. The fine lead particles removed by such work are toxic. The regulations governing the use and working of the metal must be observed at all times. The flux used may also be a cause of corrosion. Do not use more than is necessary and clean off all residues. Body solder is only suitable for body members that are known to be of conventional low carbon steel.

Plastics recognition

Plastics can be divided into two main groups, thermosetting plastic and thermoplastics. Thermosetting plastics harden under the influence of heat. Once hard, they cannot be softened or welded. Many of the components on cars are made from products in the second group, thermoplastics, and these can be softened by heat and welded. There are a number of different plastics in this group which must be identified to select the correct filler rod and temperature for the weld. The type of plastic may also affect the technique, such as the way in which the filler rod is fed into the weld.

Plastics are usually identified by code letters, often found on the reverse side of the component. To create the plastic, base polymers are frequently mixed with others which may change their characteristics. Some car makers identify these additional components and even the type of filler or reinforcement added to the base material. The base polymer indicator is usually shown first. Car maker's or welding rod supplier's data should always be used because of the many variations available. It is also imperative that the closest watch is kept on the progress of heating and fusing so that the process can be adjusted to the material being joined.

The chart below lists some of the plastics used on cars and an indication of their melting point, although these figures are given to provide general information only.

Plastics welding and bonding

Plastic components are welded together by laying a filler rod into a 'V' groove. Unlike metal welding, the heat source is a hot air gun and the filler rod used in the welding of plastics is not made completely molten. Only the outside surface of the rod and the edges of the part being welded become molten. The rod is pushed into contact with a slight but definite pressure and the molten surfaces fuse together. The temperature range when the filler and the repair material are in the correct state for fusing together is limited: too cool and the materials will not fuse together; too hot and the filler rod will be too floppy to be pressed into the groove and the materials can char.

Code	Material	Melting point (Degrees C)
ABS	Acrylonitrile Butadiene Styrene	200–285
EP	Epoxide; Epoxy resin	
PA (6, 11, 12 and 66)	Polyamide	300
PBT	Polybutylene terephthalate (linear Polyester)	220–265
PC	Polycarbonate	220–265
PE	Polyethylene	110–115
PES	Polyethersulphone	
PET	Polyethylene terephthalate	220–265
PF	Phenol-formaldehyde resin	
PMMA	Clear Acrylic or Poly (methyl methacrylate)	350
POM	Polyoxymethylene; polyformaldehyde	
PP	Polypropylene	›200
PPE	Polyphenylene ether	
PPO	Modified Polyphenylene Oxide	
PPS	Polyphenylene sulphide	
PTFE	Polytetraflouroethylene	
PVC	Polyvinyl Chloride	160 – 200
PUR	Polyurethane	
SMA	Styrene Maleic Anhydride	
TPUR	Thermoplastic Polyurethane	250–300

There are two distinct methods of plastic welding. The first is the high-speed weld using a hot-air gun with a rod feeding and laying facility. Select the correct welding tip and fit it to the welding gun. The temperature is set, the rod fed into the tip and the gun is placed onto the start of the joint. As soon as the parts to be joined and the welding rod have reached the working temperature they should fuse together immediately. Then the gun is moved backwards along the line of the repair. A steady movement will ensure that the rod will lay into the groove and fuse the surfaces together. Never stretch the rod.

Slow speed welding with hand feeding is essentially the same. Greater care is needed in ensuring that the rod is laid correctly and not stretched. The welding temperatures are quite critical. A low ambient temperature in the workshop and cold components may call for an upwards adjustment of the welding temperature. However, there is a limit to this compensation and the speed may need to be altered to avoid charring of the plastic.

As with metal welding, self-criticism is vital for continuing success in plastic welding. During welding the plastics should not discolour. The welding rod must not soften as pressure cannot be maintained. After welding, check that the rod has penetrated to the other side of the joint and formed a slight bead along its entire length. Failure of the rod to completely penetrate the joint is almost certainly a sign of a poor weld. Test by pulling the joint apart. Burnt or charred material, probably together with a failure to penetrate the joint, is also a sign of poor welding. Apart from the filling of the joint, the continuous 'seam' along each side of the joint is a fairly reliable indicator that successful welding has been achieved.

Internal vehicle components such as heaters or ducting will probably not need any further treatment. But visible items such as bumpers must be returned to their original pattern to satisfy the customer. Ingenious ways of texturing have been tried with varying degrees of success. Best practice now suggests that outside plastic trim that is painted is repaired; those that have a natural finish with texturing are renewed.

As with any other technique needing skill and practice, there are some important criteria to be observed:

- Correctly identify the plastic and the welding requirements.
- Prepare the joint as required.
- Obtain the correct size and type of welding rod.
- Tack weld the components together with a proper tip if necessary.
- Obtain the correct tip for the type of repair.
- Set the correct temperature.
- Pre-heat the joint and the rod to the correct temperature, but not for too long.
- Feed the rod at the correct angle and with enough pressure to obtain correct entry into the groove.
- Move the hot-air gun and rod at a steady speed and pressure to make a continuous, good weld.
- The indication of a good weld is an even, wavy seam of softened material along each edge to indicate a proper fusion of rod and repair and a fine bead on the underside.

Bonding systems for plastic have been introduced as an alternative to welding. They require preparation similar to that used when welding. One of these systems, for example, uses a cleaner, primer, two-pack automatically mixed adhesive, metal reinforcements and texturing material. There is a delay while the bonding cures although this can be as short as fifteen minutes if an infra-red heater is used. The degree and success of any texturing is dependent upon the creative ability of the operator. The advantage claimed for one of these materials is that it can bond all types of plastic vehicle components without the need to identify the material. Other makes available at present suggest that only the stiffer types of plastic should be bonded.

As with welding plastics, there are some important points to observe:

- The component must be clean before beginning the repair.
- Use the system cleaner before and during preparation to ensure absolute cleanliness.
- Bevel the edges to be repaired.
- Bonding usually requires an overlap so the area of the repair must be roughened.
- Any pre-bonding primer must be dry.
- Reinforcements must be incorporated before the adhesive begins to set.
- Observe any mid-repair curing requirements.
- Use the correct surface primer before finishing.

Chapter 5 | Realigning and Repairing the Body

Body Shell Alignment

The body shell and the effect of accidents

The origins and the basic construction principles of the monocoque car body were outlined at the beginning of Chapter 3. This chapter is concerned with the effect upon the bodywork of collisions or roll-over accidents. An understanding of the forces imposed by these impacts is necessary if modern body straightening equipment and techniques are to be used successfully.

Many of the terms used by experts in this field are descriptive, rather than technical (see Fig. 75). Most of them originated in America, where body measurement and straightening was first developed. In this text I have used those terms known to me. If your college or workplace use other names for the various types of damage there is no reason why you should not continue to think of them in that way. It is always important when working to written instructions, or discussing the topic with others, to clarify the slightest doubt about terminology.

Here are the major groups of damage:

■ The **Sag** is characterised by the body sagging downwards at the front bulkhead. What has actually happened is that the front of the car is pushed upward with some compression of the top and back of the engine compartment.

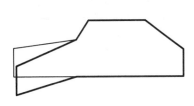

(a) The Collapse

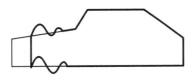

(b) The Crumple

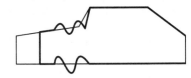

(c) The Mash

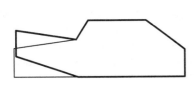

(d) The Sag

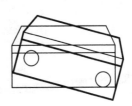

(e) The Twist

(f) The Diamond

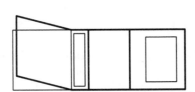

(g) The Side Sway

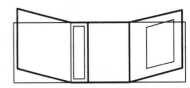

(h) The Banana

Figure 75 Body distortion by type
    (a) The Collapse
    (b) The Crumple
    (c) The Mash
    (d) The Sag
    (e) The Twist
    (f) The Diamond
    (g) The Side Sway
    (h) The Banana

■ **Collapse** refers to the condition where either the front or rear end of the car is pushed down compared to the remainder.

■ The **Twist** is where the whole body has literally twisted. Here two opposite corners will be in the same plane, while the other two are twisted onto another. As always there will be variations on a theme and it is not unusual to find one corner of the body being higher or lower than the other three.

■ The **Crumple** is the damage that occurs from frontal or rear end impact to well designed, present day cars, particularly those of European origin. It is the concertina effect produced by the progressive collapse of the front and rear deformation elements or crumple zones (see Fig. 76).
This may be right across the front or rear of the car, as frequently happens in motorway 'shunts', or the corner impact from collision with an obstacle or an approaching vehicle.

■ The **Mash** is the earlier American name given to the damage arising from this sort of impact on vehicles without deformation elements (crumple zones). In such cases the damage is likely to be more widespread and unpredictable. The passenger compartment may well be affected while the front or rear of the car could still have recognisable shape.

■ **Side Sway** is where one end of the car is out of lateral alignment with the other. The impact has moved one end bodily sideways, compared to the other. I stress this point because it is fundamental to the way the damage will be rectified. It must not be confused with the next topic.

(a)

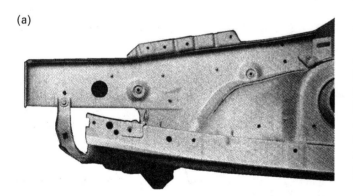

(b)

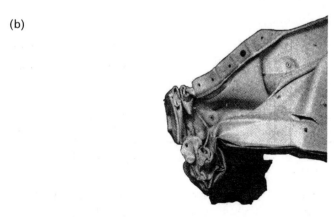

Figure 76 The deformation elements on the Volkswagen Golf are made from metal tapering in thickness; these illustrations show (a) the undamaged element and (b) after substantial impact

■ The **Banana** is the most descriptive of our names and relates to how the car looks in plan view. It is invariably caused by a side impact, usually near the centre of the passenger cell. The two ends of the car try to wrap themselves around the other object. Compared to the side sway, where only one end has moved, here both ends of the car are pointing in different directions relative to the body centre line.

■ The **Diamond** is again best seen in plan view and is the result of an impact on one side of the front or rear pushing that whole side backwards or forwards, compared to the other.

■ The **Roll-over accident**, either directly or as a result of another impact, is the remaining principal cause of damage. Damage can vary greatly, depending upon the circumstances, from relatively minor deformation of the pillars and roof to that of a total wreck.

Many impacted vehicles will not have clearly defined damage. It is more usual to have two or even more types of damage combined. A frontal impact on one side will result in the Crumple. However, in a typical frontal impact some of the forces will pass right through the structure. This is sometimes demonstrated by the bootlid opening during a frontal impact. A flexible body cell may return to the original shape after such distortion. A stiffer cell could remain deformed and a Diamond would be added to the Crumple as a result. In the case of the Banana, the impact which caused the damage will have pushed the side inwards. This has a shortening effect on the length of the car. In addition, one or both ends will have swung inwards about the centre. These are just two of many permutations and every repair is likely to be different in some way from any other.

The forces that affect the structure of a car during accidents are investigated thoroughly during the design process. Despite this, there have been many instances of permanent distortion occurring at places quite remote from the point of impact by the passage of forces through the structure. In addition, there is often secondary damage occurring after the initial impact caused by components, occupants or the load. A car leaving the road as a result of an accident will probably also sustain additional damage from striking other vehicles or obstructions. Last but by no means least is the effect of the pre-accident condition of the vehicle.

Material characteristics: relieving stress

Steel is an elastic solid and obeys Hooke's Law in the early stages of stress. That is to say that it will stretch proportionally to the load and, provided that it does not go beyond the 'elastic limit', will return to its original shape when the load is removed.

If it is stretched beyond the elastic limit, the stretch becomes permanent. A little beyond that again and the steel enters what is known as the plastic region and rapidly reaches breaking point with little or no additional stress.

All these phases will be evident in the cars that you repair. Some components of the body will have stretched and moved back, others will be permanently stretched and yet more will have failed altogether. Repair involves restoring stretched material to pre-accident shape. Steel which is torn apart must be cut out and replaced.

The steel generally used in car construction is built up of atoms of matter in cubic crystals. These crystals will lay in the steel in an ordered pattern when the body is undamaged. When the metal is distorted, and particularly if it goes beyond the elastic limit, layers of atoms become dislodged. When straightening of the components begins, the disorder amongst the atoms in the metal must be dealt with from time to time if the metal is to be coaxed back into normal shape without causing more stress. Traditionally this was accomplished by heating the metal and allowing it to cool naturally. This is not now advisable due to the advent of high strength steels (HSS). Heat can make HSS useless; they will either become too brittle or too soft.

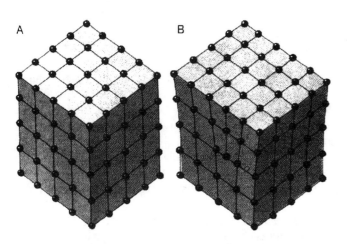

Figure 77   The crystal structure of a solid. A normal crystal is shown at *A* while the example at *B* has been damaged by an 'edge displacement' which has moved into it under stress

The alternative is to beat the metal. This has the same effect without the disadvantages. It is done at positions of likely stress, such as kinks or creases, throughout the pulling operation. Where HSS is concerned it is also vitally important that the pull or push is fully released after stress relief to check results.

The exception to these instructions concerns aluminium bodies. Heat treatment is an integral part of body construction and is also used in repair to make the parts more pliable. The temperatures used are critical and, as aluminium does not discolour to indicate temperature, thermochrome crayons must be used.

Health and safety at work and PPE

This work brings with it a number of different risks, as well as those to be expected when electrical power tools are used in close proximity to damaged metal. Here is a chart of particular hazards and the appropriate protection.

The table of health and safety hazards includes a reference to the use of correct lifting methods if back injury or even permanent disability is to be avoided. It is the most common cause of injury and, even though it is quite unnecessary, most workers accept it as a risk that must be taken.

Courses are run to provide training on lifting and you are encouraged to attend one. In the meantime always use the strength of your legs, rather than your back, when lifting heavy objects. They are a cantilever with well developed muscles. For example, in an emergency stop when driving, they are capable of exerting three times your body weight onto the brake pedal for a brief moment!

To use your legs for lifting, squat down with your back held straight, clasp the object in your hands and arms and lift by straightening the legs. If the object to be lifted is on the

Hazard	Protection
Serious injury and possibly permanent disability from the use of incorrect lifting techniques	Many of the component parts of straightening equipment and jigs are heavy; the correct lifting techniques must be used if damage to the spine is to be avoided
Heavy components falling and injuring the feet	Protective footwear to BS 1870 (200 joules) preferably with a steel midsole
Hand injury from handling heavy or sharp components	Good quality leather palm riggers' gloves
Flying particles or pressurised hydraulic oil	Goggles or face visor to BS 2092 Grade 2 or EN 166 grade F
Crushing injuries from the vehicle or equipment	Never place any part of your body between elements that could move together Always release all pulls and ensure that the vehicle is supported before attending to the equipment
Uncontrolled emissions of compressed air and pressurised hydraulic oil Playing the fool with compressed air	Examine all hoses and pipes on a regular basis Protect lines from damage, particularly jagged edges Never play with pressurised gas
Laser beams, as used on some measuring systems, may cause eye damage	When looking at laser based equipment ensure that the beam never strikes the eye Maintain all covers, shields and cut-out devices in good condition

floor it may be necessary to drag it onto a higher platform. Ideally, heavy elements of equipment should be stored in such a way that this lifting method can be used easily. Lower the weight in the same way.

There are other working practices which must be observed for your safety and the safety of others around you:

- Care must be taken that ruptured fuel or oil lines do not leak when loading damaged vehicles onto hoists or benches for the purpose of taking initial measurements . Prior to loading for repair, fuel should be emptied and any other leakage remedied.
- Do not remove the lifting appliance before the vehicle is securely clamped into position, particularly if the vehicle can overbalance.
- Mounting the vehicle and clamping it into position must be done carefully to ensure that no movement will take place during pulling.
- All fixings with securing pins or rings should be properly locked into position so as to avoid accidental movement.
- If the equipment uses wedges for location and security, check frequently that these are correctly tightened.
- Wedges and pins must be maintained in a sound, burr free condition. Where hammering home is called for, a copper mallet should always be used.
- Do not loosen or reposition mounting brackets without a hoist in position if there is the slightest danger of the vehicle toppling over. Alternatively, use a restraining chain at the end likely to lift.
- Always choose a clamp or fixings for the pull that give the highest spread of load or the most security.
- The widest possible clamping must be used where components are made of HSS.
- The vehicle attachment and the pulling equipment connection must also be protected with a properly designed safety strop (wire) at each position. This must not be so long that damage can be done before movement is restrained.
- Check continuously on the condition of the pulling attachments and the vehicle mountings. Stop and rearrange the equipment at the slightest sign of tearing or movement.
- Never stand directly in the path of the pull or anywhere along its line of action when pressure is applied.
- Ensure that no one else is in a position where there is a possibility of danger when applying the pull.
- Release tension to check and carefully reapply, making sure again that all is in order.
- Follow the equipment makers' instructions to the letter.

Equipment and tooling

Detailed instructions on using equipment are not possible, or even desirable, in a book of this nature. The makers of your particular equipment are the best qualified people to show how it should be used. What will be of use is to look at how the best can be obtained, from whatever equipment is available, by understanding the principles and practices involved. There is no doubt that a skilled professional can carry out an accurate check with a ball of string and a piece of chalk. The opposite is equally true. No amount of expen-

sive equipment will enable a raw beginner to produce a perfect job. Everyone wants a good job and there are clear moral and legal grounds for providing just that.

When the car is built in the factory, whether it be by hand or by fully automatic robots, the body components are assembled one to another with the aid of jigs. These are frames which have locating pins or brackets that exactly line up the panel with the one to which it will be welded. This assembly process continues as panel after panel is lined up on jigs and welded into place until the body is complete. This is a very accurate way of putting a car body together, but even so, the maker does not take chances and every so often a body or part body is taken aside and checked. It is placed on yet another jig and this time probes touch the body at various points and a reading is taken at each position. All this is computerised, of course, and the measurements are checked for accuracy. Such measurements can be as accurate as plus or minus 2 microns (1000 micron = 1 mm).

The same procedure is adopted in body repair except that the position of specified datum points are checked before repair to determine the extent of the damage. Then repair jigs are used to hold the components into the correct position for assembly. The equipment used for this work in the past has been of two distinct types; universal measuring systems or bracket systems.

Measuring units are basically long beams or frames with measuring scales built-in. Pointers, also with measuring scales, are mounted onto the beam or frame at various places to make contact with the body shell datum points. When the pointers are in position, the measurements can be read off and the figures checked against a chart for that vehicle. This reveals the distortion of the body.

Bracket systems are larger, heavier frames onto which are assembled special brackets to fit the vehicle under repair. Where they touch the datum points correctly the body is still true. At damaged places they either do not fit at all or there is a gap. The damage becomes obvious. These same brackets are used later to hold the new or repaired body parts in place while they are welded in position (see Figs 78 and 79).

Figure 78 A modern bench and bracket repair system with adjustable height control for the bench

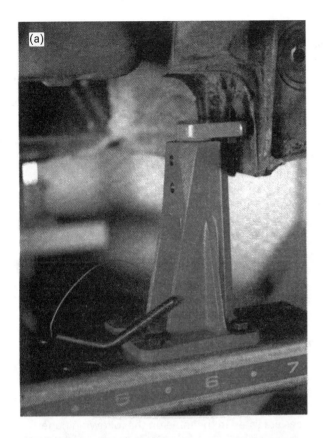

Figure 79 The Celette MZ tower and top system
In (a) the top is in the down position; in (b) the top is lifted and located by a pin

Because of the high cost of such equipment bodyshops have tended to use one or the other. However, there is now a greater awareness of the value of both methods. Measuring systems provide an undeniably quicker and more accurate

way of assessing damage in the first instance. An accuracy of half a millimetre is easily attainable. The bracket system is an all-in-one replica of the separate jigs the factory used to build the car. It provides the best conditions for accurately re-building the car to the tolerances required today. At the time of writing the two systems are moving closer together. Bracket systems are often now supplied with measurement capability and measuring systems can be provided with assembly brackets. It is unlikely that your bodyshop will possess all the different types of brackets needed for every make and model of car that they may repair. They are, however, available on hire from specialist suppliers.

Distorted bodywork is pulled out with a hydraulic ram. This is a cylinder containing a piston with a rod protruding from one end, looking rather like an overgrown bicycle pump. Oil under pressure is fed into the cylinder which pushes the piston and in turn moves the rod. Blowing into the end of a bicycle pump produces the same action. For the hydraulic ram to work it must be anchored at one end so that the extending part can exert the effort. This may be applied directly as a push but is usually needed as a pull. This pull is arranged by making the ram push a tall lever mounted on a long beam. Chains are connected between the lever and the vehicle and the other end of the beam is secured to the undamaged part of the body. When pressure is applied to the ram it will push the lever which in turn can pull out the damage. Such a device has been in use for many years, usually mounted on wheels. It is invariably called a 'Dozer®'. Correctly set-up, the dozer can handle many small repairs without the need of a jig. Used in this way it is limited to horizontal pulls but these can be done at any angle by the use of a suitable chain and body clamps to anchor the free end (see Fig. 80).

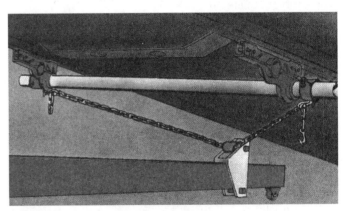

Figure 80 The Dozer® can be used to straighten minor damage all around the car if this kit is used for anchoring

The equipment used in the bodyshop for more extensive work will consist of one of the jig systems previously mentioned and a dozer or a similar device, together with many fittings. These will include additional hydraulic rams for multiple pulls, clamps for mounting the body, devices for pulling and chains. The basis of all these is a free standing frame known as a 'bench' or a frame inserted in the floor. Completely self-contained and mobile systems are now common with lifting, mounting, measuring, straightening and jig facilities all in one unit (see Fig. 81).

Figure 81 Measuring equipment can be set up to check all chassis dimensions, including the suspension strut locations, in both horizontal planes and vertically

However basic or complex the system, for it to work properly the components must be in good condition. Damaged or suspect items should not be used until repaired or verified as safe. Brackets, particularly those hired in for the job, should be checked for damage or unofficial repair. A damaged bracket used at a critical point could throw the whole job out of alignment.

Most makers of this equipment supply a great many add-on items to improve the quality of repair. These include loading devices, miniature rams for undercar use, measuring frames for suspension strut mountings and body contour frames (see Figs 82 and 83).

Figure 82 Windscreen apertures are sometimes checked with special gauges, as in the example on the Porsche

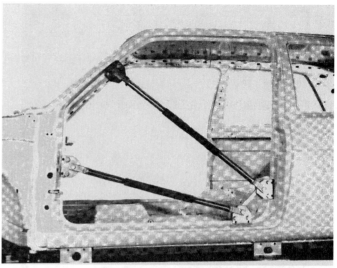

Figure 83 Door aperture bracing is used to prevent distortion during repair

Sources of information

All the makers of measuring systems provide comprehensive information and an after-sales service for vehicles not included or which give rise to problems. These and other sources are listed in the table on the following page.

Visual inspection: measuring misalignment

Before any thought can be given to the method of straightening it is imperative that the actual misalignment is clearly established and a reasonably clear idea of the cause determined. Knowledge of what actually happened to cause the damage can be most useful but is not absolutely essential. A thorough visual examination of the damaged car will normally reveal enough information to determine at least the primary collision and the consequences. Trammel gauges may sometimes be useful to confirm the general alignment of the vehicle in the horizontal plane (see Fig. 84). The final check is to accurately measure the location of the body datum points by setting the car onto a damage assessment system. The information gathered from these exercises should provide sufficient information to set up the initial hydraulic pulls with confidence.

Visual examination consists of walking round the vehicle to gain a general impression of its condition followed by a systematic and detailed assessment. This is done by working away from the point or points of impact and checking the effects. Look at the vehicle from all angles. The close angled view of panels using light reflections will reveal ripples or dents. Sometimes, feeling with the fingers can confirm a suspicion of a barely visible ripple. Such an indicator may be a valuable aid when checking the effect of the remedial pull.

At the repair stage there will already be an estimate which can be used to guide the check although it is not wise to rely totally on such documents. The inspection should be thorough. Straightening the bodywork correctly demands detailed knowledge of the vehicle as it is, and its pre-accident condition. Here are the major points to bear in mind when carrying out this survey:

Job	Source of information
Datum point positions	Datum point measurements for particular equipment Equipment makers' information system Equipment makers' enquiry service
General alignment information and dimensions	Vehicle makers' information MIRRC (Thatcham) information Equipment makers' information system Equipment makers' enquiry service
New equipment, components or accessories	Equipment makers' enquiry service Equipment makers' catalogues Trade journals Local supplier Trade exhibitions MIRRC information
Techniques of use	Equipment makers' training courses Equipment/suppliers' representative Vehicle makers' training courses MIRRC training courses Training information

- Walk round to gain a general impression of vehicle condition and the scope of the job.
- View the car from side to side and end to end, comparing outline contours and panel gaps for similarity or differences (see Fig. 85). The rear end view, for example, should show a symmetrical pattern between the roof, tailgate or bootlid and the sill panel.
- Start detailed examination at the point of impact, checking those parts of the body listed in the estimate.
- Then move to adjacent panels to see what the effects are. Are the panels in shape? Have gaps closed up? How do panels align in other planes?
- In front or rear impacts check the roof panel above the 'B' posts for signs of creasing, a good indicator of passenger cell compression.
- Contour lines are another useful indicator of alignment.
- Check the inside panels where there is access.
- The process should be repeated in an underbody examination. The straightening process usually relies on sill clamps. Are the sills in a fit state to take the load? If there is the slightest suspicion of corrosion, it must be checked.
- Much of what is seen from below will confirm the diagnosis above. Look for peculiarities that do not seem to match the damage that you are dealing with, such as an unrepaired previous impact.
- The floor pan should be examined carefully for signs of distortion.
- The underbody generally must be carefully checked for impact damage, either from the original accident or subsequent damage where the vehicle has ridden over obstructions.

Following the visual inspection, the bodywork is checked for distortion on a jig (bench and brackets) or with a measuring system (see Figs 86 and 87). Only when all this information is to hand can thought be given to the way in which the bodywork can be pulled into shape.

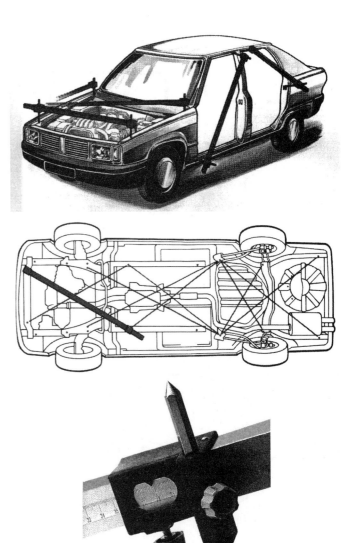

Figure 84 The trammell of years gone by is still available in modern versions for basic diagnosis

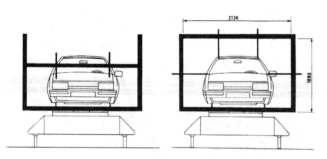

Figure 85 A body contour measuring frame

Figure 87 (a) Setting up and (b) using a measuring system during repair; note the data sheet on the car

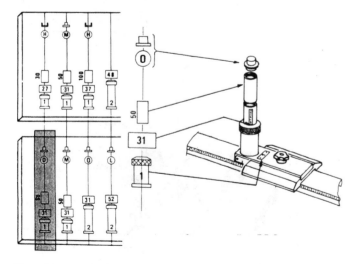

Figure 86 Part of a measuring system data sheet showing the dimensions at each point and how the apparatus is set up

Planning the pull and mounting the vehicle

Repairing body damage may be described as 'reversing the accident'. Since the introduction of the hydraulic pulling equipment that was described earlier, the industry has adopted the vector pulling principle for this work (see Fig. 88). With the undamaged part of the body securely mounted, the equipment is set-up in such a way that a pull can be made in the reverse direction to that which caused the accident. This may involve more than one pull, perhaps in different directions, to achieve the desired result. Sometimes, more than one may be needed at the same time (see Figs 89 and 90).

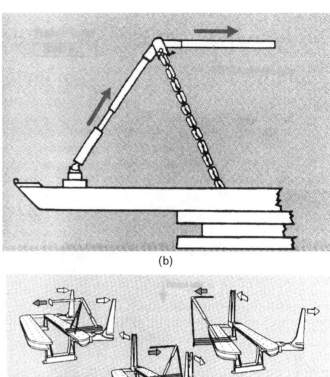

(b)

(a)

(c)

Figure 88 Vector pulling and pushing is often used to create the straightening force; (c) shows examples of multiple direct and vector pulls and pushes

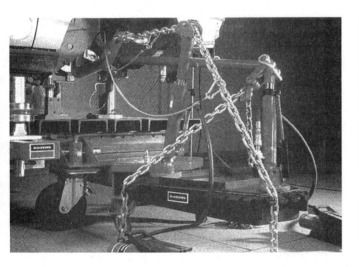

Figure 89 Combination vector pulls and pushes; note the safety strops on the pulls

Planning the pull is fairly straightforward. It is usually possible to see clearly, at least initially, which way it must go. Another factor to consider is the reaction to the pull. This is always equal and opposite in direction to the pull itself, although it may be shared through several mounting points on the jig. This must be carefully assessed if the vehicle is to remain on its mountings and not be damaged in another way.

Here are some typical examples of both the pull and the reaction anchoring:

■ The condition called Sag calls for the body to be held down at the 'best end', with supports placed under the limit of the straight part of the body and a downward pull applied at the lifted end. Care must be taken to see that the pull is directly downward or a Side Sway may be induced.
■ The Collapse is the reverse of the Sag. The body is held down close to the point of the bend, the 'best end' is supported and an upward pull applied at the end of the collapse. Once again, the pull must be vertical if Side Sway is not to be induced.

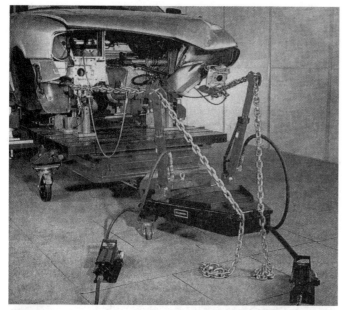

Figure 90 A double pull and a downward pull

If, in both these instances, measurement shows that there is also a Sideway element in the damage, then the pull can be angled to remove this at the same time as the main pull is applied:

- A Twist is removed by clamping the body across one end, supporting the lower corner at the other end and pulling down on the high corner.
- Crumple or Mash damage involves a straight pull with the other end of the vehicle firmly clamped. The forward end is supported just behind the damaged section.
- Sideway rectification is simple and straightforward, provided that there are no other complications. It is necessary to hold the main part of the vehicle and pull over the end that has moved sideways. There is the possibility of inducing a Sag or a Collapse during the pull, so support should be provided at the limit of the damage.
- At the other end of the scale the Banana is one of the most complex jobs as far as setting up the equipment is

concerned. To reverse the effect, the ends must be pulled sideways to their original position, at the same time as the car is tensioned lengthways to pull it out to length. Then, at or near the point of impact, a pull or push must be applied to reverse the initial impact. It takes special care and not a little expertise to control all four movements to give the desired result.

- The Diamond calls for a little extra care. One side of the car is firmly clamped and a pull applied to the backward corner to pull it forward. It is most important that the pull is angled along the line of the diagonal and a reaction anchor be provided at the opposite corner, also aligned to the diagonal.
- Where Roll-over damage is concerned this is likely to be rather more gradual, with pulls or pushes to draw out the roof and pillars as needed.

The one thing that is not done in repair is to copy the speed of the deforming process; the pulling is carried out carefully and slowly with many stops for relieving stress.

The details of mounting a vehicle to the straightening equipment will depend upon the type to be used as well as the nature of the pulling. There are, however, some general guidelines which can apply to most work of this nature:

- Use the equipment data sheet.
- Contact the company's technical support section if there is any difficulty in applying the data to the vehicle under repair.
- Observe safety rules, use the correct straightening equipment and suitable PPE.
- Lift heavy components properly; seek help if necessary.
- Work in a sequence from undamaged to damaged.
- Select at least four datum points on the undamaged part of the vehicle.
- These reference datum points should be as far apart as possible.
- Where there is a choice of attaching adapters to either vehicle or equipment, choose the vehicle.
- Adapters, for either measuring systems or bracket benches, must fit correctly to the position.
- Where brackets are used for reference, there must be no strain or interference of any sort.
- Check and check again. Re-check the datum and other mounting points during the mounting process.
- Where final tightening of items such as sill clamps is supposed to be done in a number of stages, make sure that this is completed before the pull is applied.
- Do not remove any lifting appliance or holding straps until the vehicle is secure.

Pulling out the damage and verifying alignment

Present day equipment permits vector pulls at any angle and, as discussed earlier in this chapter, anchoring to cope with the reaction will have been considered before mounting the vehicle. It is essential that the parts of the body structure to be anchored are firmly secured and any reaction anchor is securely fitted (see Fig. 91). There should always be an anchor or support as close as possible to the beginning of any damaged section.

Figure 91 Straightening demands adequate anchoring of the body, as (c); two types of anchor for the opposite reaction forces are shown in (a) and (b)

Once the vehicle is mounted accurately and securely, the components can be set-up for the first pull or pulls. The equipment supplied for this part of the work may vary in detail but, in general, the components are very similar. Any specific recommendations made by the makers must, of course, be followed. Here is some general guidance on setting up the pull, operating the equipment and checking the alignment.

Take time in considering how best to attach the pulling attachments to the vehicle. This is important for a number of reasons. The very success of the pulling operation depends upon a good grip of the damaged elements. Taking trouble to obtain a good attachment for the pull in the first place will avoid wasting time in finding another location later. The risk of personal injury is also greatly reduced. Spread the strain as much as possible, particularly where HSS material is involved.

When a good attachment for the pull has been arranged, attach a chain. Draw it out away from the body in the direction of the pull. Put yourself into a position where you can look along it to obtain a view of what may happen as the strain is applied. Consider, too, the anchors that have been provided against the reaction. Are they in the right place? Once certain of the direction of pull, the equipment can be positioned and set-up to provide the effort.

All straightening equipment has the means, through various attachments, of applying the effort in any direction. Where it is being redirected by means of posts or pulleys, make sure by trial application of a little pressure, that what you expect to happen can take place. It often helps where these complicated set-ups are involved to draw them out on paper just to make sure. Again, the reaction is vitally important if more damage is not to be created.

Some of the bodywork will almost certainly be cut away and be replaced. It is often the case, however, that the pulling chains or fittings may be pressing against components that are reusable. Look out for such instances and protect these body parts with the soft blocks that are available or with cloth padding. A few blocks of wood may also be very useful for such purposes.

When the pull is ready, attach as many safety strops (wires) as necessary to hold the components should any slip or fail. The strops should have enough freedom to permit the movement that is expected but not so much that a flying component could strike anyone or anything. When the pull is under full pressure, a blow from a released component could easily be fatal. It follows that no other person should be close to the job at this time.

The signs of deformed bodywork, such as closed-up door gaps or roof creases, should be kept in mind as reference points to assess the effect of the pull. It is helpful to have as many widely distanced points as possible. As the pull is applied they must all be monitored on a regular basis. It is usual to have a fairly rapid initial movement of well collapsed body members such as occur with Crumples. Once this stage is reached or where there is only a small movement, the pull must be stopped and progress checked. Where body members are under stress, as occurs with a crease being pulled out, the stress must be relieved by beating. If the material is thought to be HSS, the pull should be released momentarily and then reapplied.

It is also important to monitor the general vehicle condition, particularly at the points where the reaction anchor is situated. If the pull is having the intended effect without adverse consequences it may continue without alteration. At the first sign of any adverse effect the pull must be stopped and reconsidered. This may mean a slight adjustment or a total rearrangement.

The straightening process continues until as much of the damaged section of the vehicle as possible is repositioned and is free of tension. Now a thorough check is made of all datum points to ensure the best basis for the repair stage. Where the vehicle manufacturer's data provides alignment figures, it may be considered prudent to use these as a cross-check.

Preparing the Body for New Components

Health and safety at work and PPE

Many of the precautions relating to straightening mentioned earlier also apply to the repair phase. In particular, the centre

of gravity of the vehicle may change quite dramatically if large parts of the structure or heavy components are removed. Additional supports or restraints may be needed, especially if mountings are to be slackened or released.

Sources of information

It is always important to check out details of the repair process before commencing cutting away damaged body members and panels. Makers of both genuine and replacement parts have a habit of changing the design of components and their fittings. Part panels or a combination of sections may be supplied instead of an assembly. Sometimes original components must be salvaged for reuse or new parts fabricated. If there is any doubt about the assembly or the process, contact the supplier for advice.

Equipment and tooling

Spanners and sockets are needed for bolted fastenings and drilling or cutting equipment for welded or bonded components. Damaged bolts can be freed with nut splitters, sawn off or drilled away. Where there is a layer of vinyl or similar

Hazard	Protection
Jagged metal and broken glass	Riggers' gloves with leather palms Strong footwear to BS 1870 (200 joules) and steel midsole
Flying particles from cutting and grinding equipment	Goggles or a visor to BS2092 Grade 2 or EN 166 grade 'F' Industrial overalls
Excessive noise	Ear defenders as appropriate Warn others within hearing distance or use screening
Back injuries from incorrect lifting procedures	Follow the correct procedures for lifting heavy or awkward objects
Fire risk from oil and fuel residues	Pre-repair cleaning and thorough draining of systems likely to be affected Revise fire precautions
Obstruction of gangways or work areas with components and scrap	Place removed components in a designated area or remove to scrap bin
Damage to repaired vehicle and others nearby from metal and other particles	Protect undamaged parts of the vehicle under repair Protect vehicles nearby or move to a safe distance

Job	Source of information
Dismantling procedures	Vehicle makers' information (manuals, service letters, training information) MIRRC (Thatcham) information Proprietary manuals
Cutting lines	Body dismantling techniques Vehicle makers' information (manuals, service letters, parts information, training information) MIRRC information Equipment makers' information Computer estimating systems

protective, this can be cut away with a sharp knife. It sometimes helps to heat it with a hot air gun. The usual hand cutting tools will always be needed from time to time, together with a good hammer, a bolster and cold chisels.

Cutting should be done as cleanly as possible to minimise preparation time for reassembly. Panels should be cut with a clean cutting tool such as an oscillating saw, preferably pneumatic. Rotary cutting discs may also be used. Folded edges are ground away along the edge. A belt sander is invaluable for gaining access to tight corners, particularly where brazing must be removed. A mule skinner, a type of rotary wire brush embedded in resin, is good for removing paint finish to find spot welds. Use a drilling jig whenever possible when drilling out spot welds as they have a stop which can be set to prevent drilling through the under panel. A flange setting tool will be useful for preparing the joints of some part panels. A set of sharp drill bits should also be available.

Removal of damaged components: cutting lines

A clear idea of which components need to be removed and where they should be cut back is essential (see Fig. 92). Any supplementary cuts to gain access to awkward areas should also be clearly worked out. Use the vehicle maker's information whenever possible, in conjunction with that from the MIRRC. Some computerised estimating systems also include this information. In any event, always compare the replacement components to the work to make sure that there are no changes.

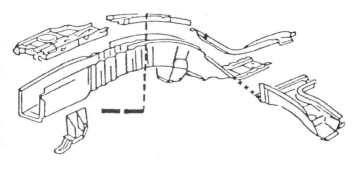

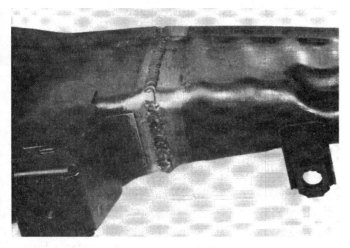

Figure 92 (a) A car maker's recommended cutting line for deformation element, and (b) how it looks after repair

It is sometimes possible to carry out the repair in an easier way by using only part of a new component. This method is usually acceptable provided that the cut is clear of critical locations, such as mounting points, and that the welding process and joint that you propose to use is adequate for the situation and the expected loading.

The techniques to be used during this work have been discussed in previous chapters. It is worth mentioning again that the stripping down should be done carefully, especially if the vehicle is complete. Some components will probably be badly damaged but even so a little extra care taken now can save time and effort later. A wiring harness, for example, may be trapped but otherwise undamaged. Some care taken to release it could avoid damage resulting in costly renewal. Checking damaged components against the estimate will ensure as far as possible that replacements have been ordered. It is also wise to look for special brackets or mountings that are attached to others and could be separate parts.

Repair and preparation for rebuilding

Once again, a general check of the repair is a good idea once the main pulling and dismantling has been done. The upper parts of the body will have been visually assessed or measured for position during the pulling and the alignment may be correct. However, it is the visual appearance that is seen by the customer and that should be checked continuously from now on. Although the body is in alignment there may be some finer adjustments to make to door openings or other apertures. Look at the new parts and compare them to the area under repair. Do the doors have even gaps around them? Is the 'A' post to door gap the same as the gap at the 'B' post? Is there a smooth flowing shape from one panel to another? All these, and as many others as you can think of, are visual indicators of a good job.

Once the bolted components have been removed the bulk of the damaged material can be cut away. If any of the remaining components need reshaping, it is best to do this by comparing them against new adjoining parts, perhaps clamped into position on the jig brackets. In this way accuracy can be guaranteed and unnecessary work avoided. Always remember that you can easily cut more away; it is often more difficult to put it back! Apart from this work, the remaining joints can be cleaned off in preparation for applying a welding primer.

As always, clean, corrosion free metal is necessary for successful bonding or welding. It is also essential that the metal is not reduced in thickness in load bearing or safety elements of the structure by grinding or sanding. Rotary wire brushes or a mule skinner will usually achieve the desired result. This is also an important point to remember where abrasives are used to cut back the remains of welds or other excess metal. With material thicknesses of around 1 mm, the slightest error with a coarse abrasive can reduce the thickness dramatically. Wherever possible, use lighter, gentler equipment and less coarse abrasives for more sensitive work. The belt sander is ideal for many of these jobs and has the advantage that it will get into awkward corners easily.

Where part panels are being used, overlapped joints may need to be prepared. MIG plug welding calls for the appropriate size of hole to be punched into the outer panel (see Fig. 93). Heavy materials, such as those used in sub-frame members, will need grinding to an angled edge.

The utmost care must be taken in the preparation of long-itudinal members or any component that provides location for suspension or steering assemblies. Most modern vehi-

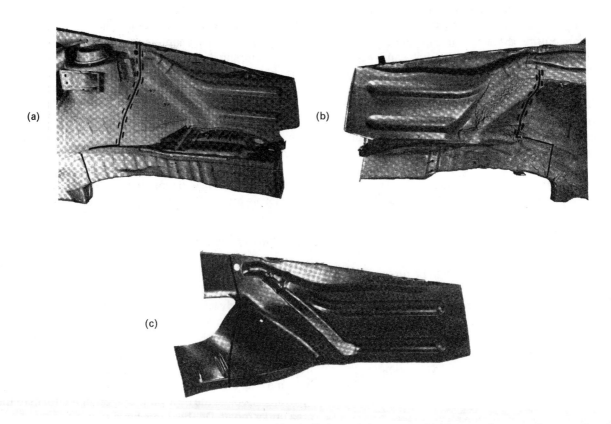

Figure 93 (a), (b) Typical cutting lines on the inside and the outside of a flitch panel, the new panel (c) is cut to give an overlap as can be seen from the position of the hole at the top

Hazard	Protection
Sharp edges of metal panels Heavy components	Footwear to BS 1870 (200 joules) with steel midsole
Flying particles	Goggles or visor to BS 2092 Grade 2 or EN 166 grade 'F' Riggers' gloves with leather palms Industrial overalls
Noise from cutting, drilling and grinding equipment Noise from beating and hammering of a continuous nature	Ear defenders to BS 6344 Part 1 Warnings to others in vicinity or noise screening
Noxious fumes from welding Vapours from bonding agents and other chemicals	Face masks to BS 6016 or EN 144 FFP 2S Use Grade 3 for MIG welding to protect from ozone (all metals) and zinc oxides (galvanised steel)
Dust and particles from grinding and sanding panels and filler	Masks to BS 6016, HSE approved for nuisance dusts or EN 149 FFP 1
Dust and fumes in the work area at large	General extraction should be used at all times to avoid con-taminating the workplace
Lifting heavy or awkward objects	Always lift with the legs; keep the back straight Seek help for items which must be held at a distance from the body

cles are provided with limited adjustment for their steering geometry. Indeed, it is common to find vehicles that do not have any adjustment at all for some of these settings. The only way that these sensitive assemblies can be properly aligned is to repair the body as accurately as it was built in manufacture. The newest designs are calling for accuracy in the order of 1.5 mm. That demands very great care in preparing the component, as well as in setting it up and fastening it into place.

Aligning and Attaching Body Sections

Health and safety at work and PPE

The rules on handling heavy weights should be carefully observed. It is also important to remember that holding or moving lighter weights at some distance from the body may cause as much injury as those that are heavier. Lifting items such as bonnets into position can be risky without help. Chemical materials used for anti-corrosion treatments, primers and sealers all have specific health and safety requirements which must be observed.

Sources of information

General information relating to some of these activities can be found in trade magazines from time to time.

Mounting and aligning components

There are three main considerations when mounting components. Are they in the correct position, can they be properly fastened in place and do they look right (see Fig. 94)? The last one, although the least important technically, is the only way in which the customer can judge your work. Meeting all these requirements demands accurate and careful assembly onto correct and undamaged jig brackets (see Figs. 95 and 96). This is of particular concern where such

heavy components as the engine and transmission have been removed for repair work. The load imposed by these components will alter the shape of the body to unacceptable limits unless the alignment is absolutely accurate and without pre-stress.

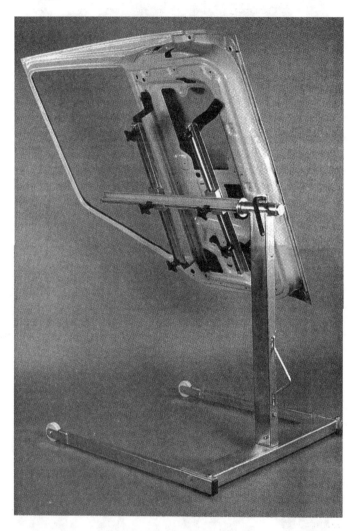

Figure 94 Working at the correct height and with components properly fastened makes for fast, safe work; this door skinning rig is one of many devices available

Job	Source of information
Aligning panels and components Checking body dimensions Mounting components into position	Vehicle makers' information (manuals, service letters and training information) MIRRC (Thatcham) information Equipment makers' data sheets
Welding components into place	Vehicle makers' information (manuals, service letters and training information) MIRRC manuals and training information Welding equipment makers' or suppliers' manuals
Anti-corrosion measures Priming and sealing	Vehicle makers' information (manuals, parts literature, service letters and training information) MIRRC manuals and training information Material makers' and suppliers' manuals

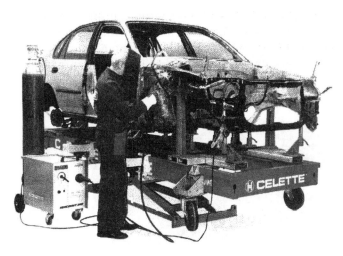

Figure 95 Repair under way with the new components located by the brackets and tops

Jobs that only need cosmetic panel replacement often have the new panels clamped into a position that looks right. Some manufacturers provide detailed settings and tolerances for all the panel gaps on their vehicles. Use the inspection technique covered earlier in this chapter. The easiest way of mounting new components for more extensive repairs, particularly where important datum points need aligning, is to use a bracket system. Bodywork can also be assembled using a measuring system. This method calls for a little more care in clamping the components together.

Figure 96 MZ towers and tops holding both a bodyshell and new components in place; a telescopic gauge is being used to check other measurements by comparison

Whichever equipment is used, the location or measuring point must be absolutely accurate. For those using a bracket system it is essential to make sure that the component is not under tension when it is held in place by the bracket. Even though other components may be welded onto it subse-

quently, the inaccuracy will remain and may become worse in use. Always check, and check again. The only acceptable accuracy is no tension or, in the case of a measuring system, within the tolerance allowed.

Be on the look out for steering or suspension mountings that could be out of position, and particularly where the steering rack is mounted on the body. This can cause the problem known as 'bump steer'. Whenever the suspension flexes the road wheels swing to one side, although the steering wheel is held still. All steering and suspension positions mounted on the bodywork must be aligned to the vehicle centre line. If no data is available make them equidistant from the centre and in height, and level in the longitudinal plane.

Another topic, which is not always agreed upon by everyone, is how much loose assembly should be done to check how the new components will go together. Although it takes time there are a number of advantages. The sequence of assembly can be rehearsed. Any remaining misalignment in the body should become apparent as the components are assembled. It also provides a check on the appearance of the finished job. Any major problems with visual alignment should become apparent.

Once the assembly is accurately and visually aligned it is often helpful to mark some of the more adjustable positions. This can cut down on the number of times that components have to be checked again for correct positioning. Nonetheless, your motto should always be 'if in doubt; check it out!'

Component attachment: verifying alignment

For details on attachment by welding, brazing, bonding or mechanical fastening refer again to Chapter 3, 'Replacing Body Panels and Fittings'. The important aspects to stress here are those that concern the attachment of components which contribute to the strength and safety of the passenger safety cell and the crumple zones. Be guided by the makers' information or that issued by MIRRC at Thatcham. Here are some reminders:

- Use only the number of spot welds or MIG plug welds given in the vehicle data.
- It is good practice, and essential for some makes, to use a zinc-rich weld-through primer on flanges before assembly.
- Check and re-check alignment as components are assembled, particularly where there are long runs of MIG weld.
- Remember always that the job must be visually acceptable; take care over the alignment of one panel with another.

Anti-corrosion treatment, priming and sealing

At first glance these topics may seem to be out of sequence. This is not so because anti-corrosion starts with the very foundations of the job. Chapter 3 dealt with the protection that should be given to the job as work progressed. Amongst the points raised was that of preventing swarf and other debris from dropping into open box sections. Such material can also collect in crevices in the bodywork as the job progresses.

The first step in anti-corrosion treatment then is cleanliness. Inner cleanliness is absolutely essential if metal particles are not to corrode inside box sections or wherever they may be trapped. It must be remembered that moisture laden air is present inside cavities, just like that on the outside. Whenever the temperature of the metal falls, moisture condenses onto the inside surface. Rain water draining from the windows will also be present inside the doors in some quantity. Their internal shape too, is such that they can easily hold trapped particles.

If material such as swarf cannot be removed from an enclosed section, as a last resort it may be flooded with enough preservative wax to prevent its exposure to the moisture laden atmosphere. Because of the presence of moisture, drainage openings are provided in the bottom of such cavities. These must be retained during rebuilding and must not be blocked by wax.

Anti-corrosion is also the reason for applying zinc-rich weld-through primer to weld flanges before assembly. The pre-painting of some inner panels may also be called for in this respect. Yet another element in anti-corrosion treatment is the use of an excess of adhesive sealer when bonding components together. This prevents air pockets and reduces the possibility of corrosion within the joint.

Some car makers also use sacrificial zinc anodes on bolted body fittings. They are designed to gradually waste away instead of the steel body corroding, a lesson learned from the marine industry. They should always be renewed when assembling such components. Be careful that the makers' information is followed. Zinc anodes must sometimes be fitted in repair when they were not used in original manufacture.

Once any filling is completed and the bodywork has the correct contours, priming must be applied before sealing can be done. I have assumed that most bodyshops will ask the painter to apply the primer, so details of the process are given in Chapter 6.

When the primer has cured, sealing of the seams can be carried out (see Fig. 97). During production of the new car, sealing is applied from guns continuously fed with sealer. The workers who apply the sealers do so for much of the time and develop individual skills in doing so. The aim in the repair shop is to produce, as nearly as possible, the appearance of the original product. You may not know that a particular shape is made by a quick wipe with a thumb (as was mentioned in Chapter 3) but in many instances the factory finish can be reproduced. Many makers of sealing materials now provide sealing guns and cartridges that will produce near exact copies of the original material if a suitable technique is used.

Care must be exercised in choosing a sealing product, although this is a decision probably made for you. If you do have a choice, bear in mind that if there is an anti-corrosion warranty on the car, the maker of the car may specify which sealer is used, and where. Some makers supply an extensive range of sealers for use in many differing locations. Use of an unapproved sealer could invalidate the warranty.

Sealers must be applied to clean, dry surfaces. If the bodywork is cold, the atmosphere need only be a little warmer for condensation to occur. This will not only trap moisture under the bead, it will probably lead to failure of the seal. Corrosion is almost inevitable. If in doubt use a hot air blower or a hand held infra-red heater to warm the seam and dry any moisture before sealing. Observe any requirements regarding temperature of the sealant, too. They may usually be warmed to aid application.

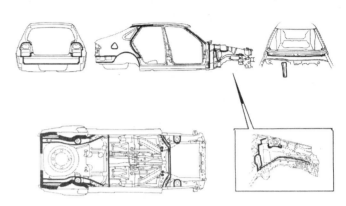

Figure 97   Some of the areas where sealing is required on the Saab 900

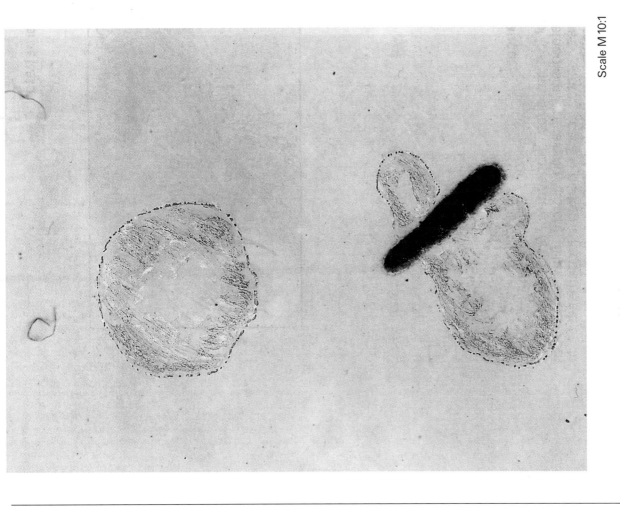

Plate 2 Damage caused by bee excrement

Droppings from healthy bees are a few millimetres long and about 1 mm across. In shape they resemble a comma and are brownish yellow in colour. Droppings from unhealthy bees are circular, 3–4 mm across and yellowish brown.

Plate 1 Damage caused by aphid excrement

In the initial stages, large yellow droppings approx. 1 mm across, which leave behind small dull marks after being cleaned off. Following this the paintwork can be eaten into to the depth of the filler, leaving a circular erosion with a central 'island'.

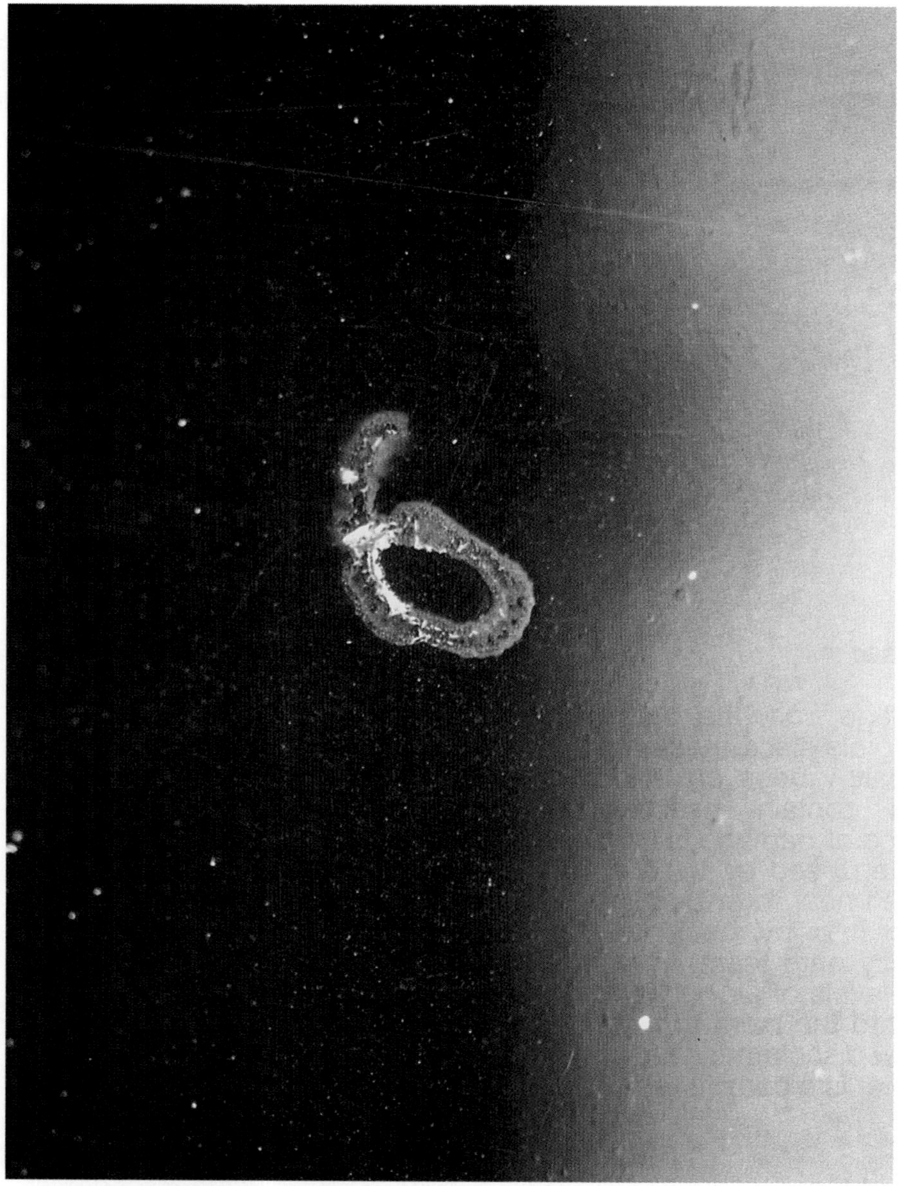

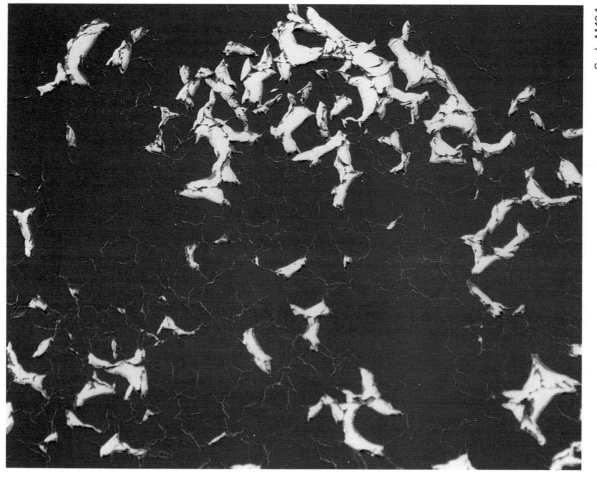

Scale M 10:1

Plate 4 **Damage caused by bird droppings**

Dull marks on the paint topcoat with initial or advanced signs of corrosion and in part web-like cracking of the paint surface. The appearance of the damaged areas resembles the effect of caustic substances.

Scale M 10:1

Plate 3 **Damage caused by flies, mosquitoes, thunderflies**

Initial or advanced corrosion of paint topcoat. In some cases filler may be visible. The corroded marks resemble a letter G, C, U, or O.

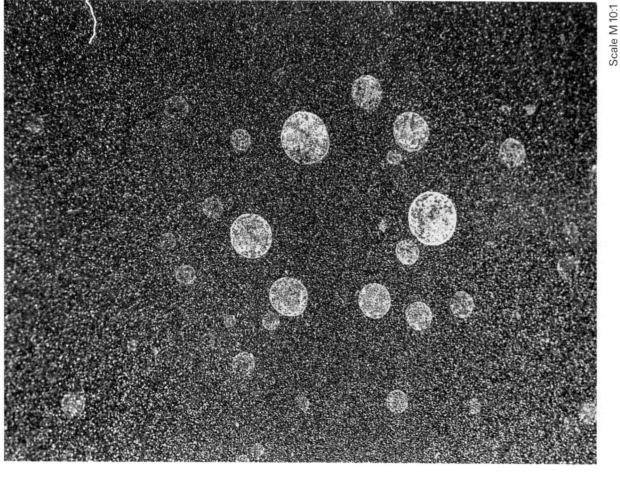

***Plate* 6   Damage caused by brake fluid**

Brake fluid causes corrosion and blistering. If it is allowed to interact with the paint for long periods, filler and primer are also damaged.

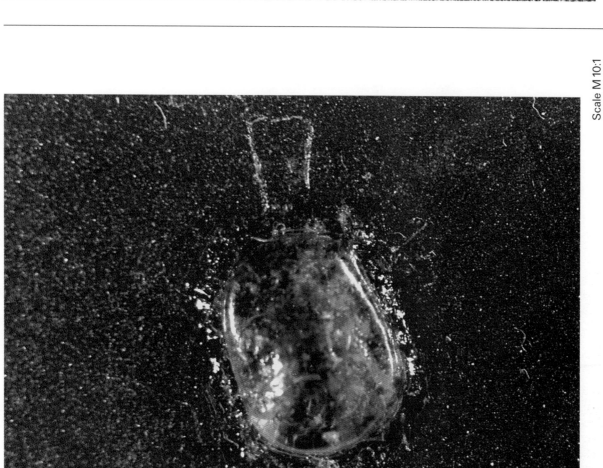

***Plate* 5   Damage caused by tree resins**

When trees are in blossom or if foliage or branches are damaged, resin is exuded and drips on to vehicles parked underneath. If the drops of resin are not removed immediately, they cause brownish-yellow discolourations.

Plate 8   **Damage caused by sulphuric acid (soot)**

Oil-burning installations (power stations, industrial furnaces, heating installations for private households, etc.) release sulphur and soot into the atmosphere. Combined with humid air rain, these substances form sulphurous acid and sulphuric acid. Raindrops then leave greyish-black marks up to 2 mm across on our vehicles.

Plate 7   **Damage caused by polishing**

Circular patches or rings of varying sizes with reduced surface gloss and smooth paint surface.

image 1:1

Plate 10 Loss of gloss (sinking, poor hold out)

The topcoat surface appears uneven and with a slight texture.
The micro-structure of the surface can give rise to a reduction of gloss.

Causes

1. Insufficient curing of the primer before application of the topcoat, the drying time was too short, or the film build too high.
2. Coarse filler was applied without the application over the top with a fine stopper.
3. Too few coats of topcoats applied.
4. Topcoat applied too thickly. Residual solvents can only escape after a long drying period leading to sinkage.

image 1:1

Plate 9 Floating (mottling, flooding, clouding)

The surface of the topcoat is uneven in colour.

Causes

Floating of a topcoat can be caused by the following:

1. An extremely thick layer of topcoat.
2. Too high or too low spray pressure.
3. The use of incorrect thinners.

5

image 1:1

Plate 12 Poor hiding (poor covering, poor opacity)

Old paint, spot primer or areas of filler are visible through the topcoat. The paintwork is 'dappled' and uneven in colour.

Causes

1. The substrate was not even in colour all over the painted area.
2. The topcoat was not thoroughly mixed before use.
3. The incorrect thinner was used.
4. The coats of paint were too thinly applied.

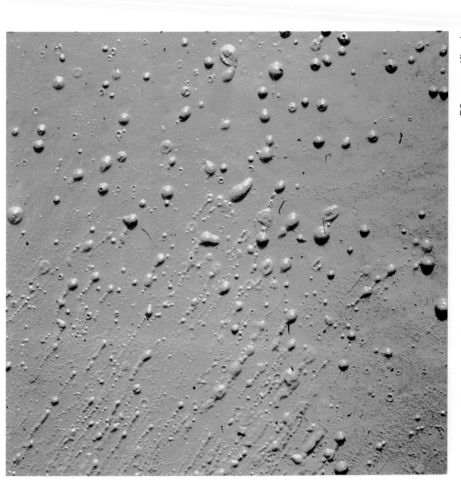

50 x magnification

Plate 11 Blistering

Moisture blisters can develop in various forms, sizes, areas and density.
Blistering can occur between the individual layers and also beneath the entire paint structure. The paint film is enclosed so that the blisters disappear in dry weather.

Causes

1. The surfaces to be coated (filler, bare metal, etc.) have not been adequately cleaned. Contamination from soluble salts in water due to dirty water used for sanding or cleaning or sweat from hands (wipe marks like a 'string of pearls' with a clearly visible arrangement of the blisters) and subsequent effects of moisture on the newly refinished coating over a long period of time.
2. Mechanical damage to the protective coat of paint and subsequent infiltration of the coating round the damaged area.
3. Wet sanding of polyester materials without allowing sufficient water evaporation times before processing with primer and topcoat materials.

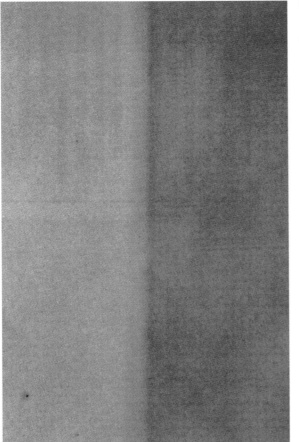

Plate 13   Bleeding

Bleeding is a discolouration of the topcoat often occurring as a red or yellow colour shadowing.

Causes

1. Soluble pigments (dyes) from the original coating are dissolved in the solvent of the repair materials and discolour the surface.
2. Bleeding through can also occur when excessive peroxide from the polyester filler is dissolved by the solvent in the repair material. The peroxide then reacts with pigments and there is a yellowy-brown discolouration in the filler spots. Blue and green tones are particularly prone to this.
3. Bitumen or tar residues.

Plate 14 Industrial fall out (contamination, staining, lime marking)

The paintwork has been attacked and corroded or discoloured by aggressive substances such as industrial waste gases, resins, petrol or chemicals, which in the worst cases cause the surface finish to be destroyed.

Causes

1. In the case of discolouration by tar, components migrate into the paint surface and dirty brownish black specks are left.

2. Industrial waste gases, chemicals or tar penetrate into the surface of the paint and discolour the topcoat. This can be caused by a chemical reaction with the pigments (e.g. in the case of acids).

3. Aggressive substances such as resins, petrol and bird lime corrode the paintwork. Depending on the length of time the substance is left to react, the corrosive effects can be so bad that the paint is decomposed and therefore destroyed.

image 1:1

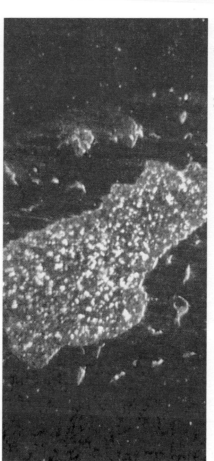

60 x magnification

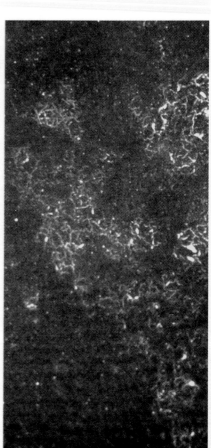

60 x magnification

8

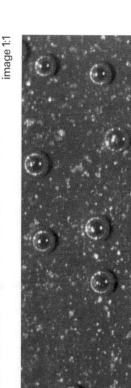

image 1:1

60 x magnification

Plate 16   Boiling (blowing, popping)

In the case of this surface problem blisters occur with a fine hole in the middle.

Causes

The occurrence of these bubbles can have various different causes:

1. The coats were applied too thick or heavy.
2. The use of unsuitable thinner (too fast).
3. Too short a drying time between the individual spray processes.
4. Too high an object temperature to accelerate the drying.
5. Too long an air drying time for two component paints before the object is placed in the oven
6. The use of infra-red lamps can cause too high a surface temperature if the lamps are close to the object.

100 x magnification

Plate 15   Poor adhesion (peeling, flaking)

Adhesion loss can occur in two different ways. Firstly there can be adhesion problems to the substrate (total paint structure), and secondly there can be an inadequate bond between the individual coats (intercoat adhesion).

Causes

Adhesion loss can occur if:

1. Substances which cause adhesion problems are left on the substrate to be coated (e.g. silicone, oil, fat, wax, rust, polishing residue, etc.).
2. An incorrect primer was applied to the substrate.
3. Sanding of the substrate was inadequate or not carried out at all.
4. The primer or basecoat was too dry or too thinly applied.
5. Flash-off time between coats of basecoat metallic are too short with the material being applied too thick.

9

image 1:1

Plate 18 Runs (sags, curtains)

These are beads, droplets or even large globules (so-called 'curtain-effect') which have run vertically down the panel.

Causes

1. The nozzle on the spraygun was too large.
2. The spray technique was not suitable for the material. The material application was too wet because the gun was too close to the object or the spray movement was too slow.
3. The coats of paint were applied too thickly.
4. The flash-off time between individual coats was not long enough.
5. Thinner or hardener too strong for the particular spray conditions. This happens particularly when spraying small areas or in low spray temperatures.

100 x magnification

Plate 17 Cratering (fish eyes, cissing)

Craters are circular dents with raised edges in the topcoat or the intermediate coats.

Causes

1. Oil, fat, wax and silicone polish residue were not thoroughly removed from the surface to be coated.
2. Contamination from the air, e.g. spray mist from another type of paint.
3. Oils or water from the compressed air.
4. Silicones from the aerosol cans (water repellants).
5. Foreign substances from industrial plants nearby.

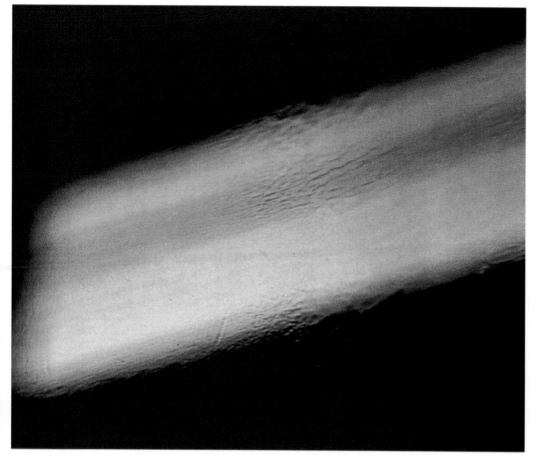

image 1:1

Plate 20 **Lifting (sweating, swelling)**

Marks showing the definition of edges of old paint, or primer or spots of body filler or stopper.

Causes

1. The old paintwork to be refinished was not sanded finely enough.
2. The feathered area of primer to old paint or of old paint to the body metal was too coarse and was not sanded down as carefully as was necessary.
3. The fillerspots were not thoroughly and carefully sanded down.
4. The filler was sprayed too thick and not allowed to dry sufficiently.

image 1:1

Plate 19 **Orange peel (dry spray, poor flow)**

The paint has an uneven texture which is similar to the skin of an orange.

Causes

1. The spraygun is held too far from the surface being sprayed.
2. The spray pressure is too low and the atomisation not fine enough.
3. The topcoat is applied too lightly.
4. The spray application is too dry because the paint supply from the spraygun is set too low.
5. Viscosity of the paint is too high.
6. Thinner used is too fast.
7. Surface or spray temperature.

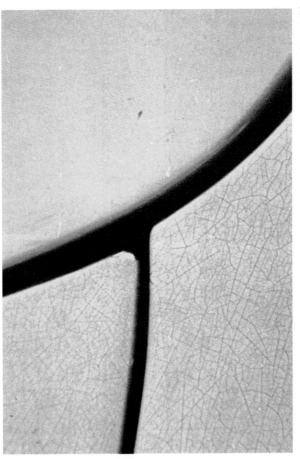

image 1:1

Plate 21 Cracking (alligatoring, crocodiling, crazing)

Cracks of different length and width spread in various directions in the top coats.

Causes

Normally cracking occurs due to wide fluctuation of temperature acting on the paint film build under the following conditions.

1. When the paint film build has too high a coat thickness.
2. Application of a repair finish over old paintwork which already contains barely visible hairline cracks.
3. Use of paint materials which are not designed to adapt to each other's hardness or flexibility (e.g. a hard and rigid polyester stopper or filler, applied to a thermoplastic acrylic T.P.A. topcoat, will lead to cracking due to the tensions caused by different expansion and contraction forces.)

400 x magnification

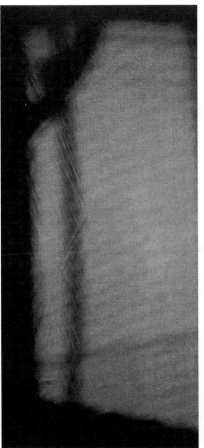

image 1:1

60 x magnification

Plate 23 Sanding scratches (sanding marks)

Sanding marks are visible either individually or in a large number as grooves in the surface of the paint which follow the lines of sanding operations prior to painting.

Causes

1. The primer and/or the filler was sanded with paper that was too coarse. The sanding marks then show up in the next coat of paint as small grooves in the surface of the paint.
2. The recommended drying times of primer were not adhered to. The sanding marks become visible after the drying of the paint finish sinks into the uncured primer material.
3. The use of sanding discs and papers on the car body which are too coarse always increases the possibility of the appearance of sanding marks.

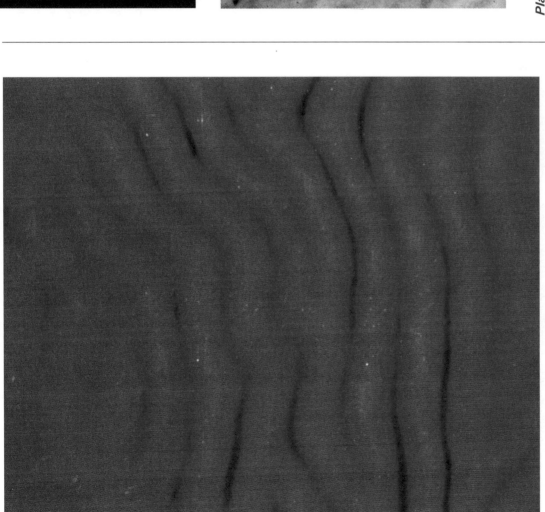

25 x magnification

Plate 22 Wrinkling (crinkling, puckering, rivelling, shrivelling)

Irregular grooves/ridges form on the surface if the surface layers of paint dry much quicker than the layers of paint below. The surface of the paint then 'wrinkles.' This only happens with synthetic paints which dry by oxidation.

Causes

1. The synthetic topcoats were applied too thickly.
2. Unfavourable drying conditions (e.g. very high room temperature).

13

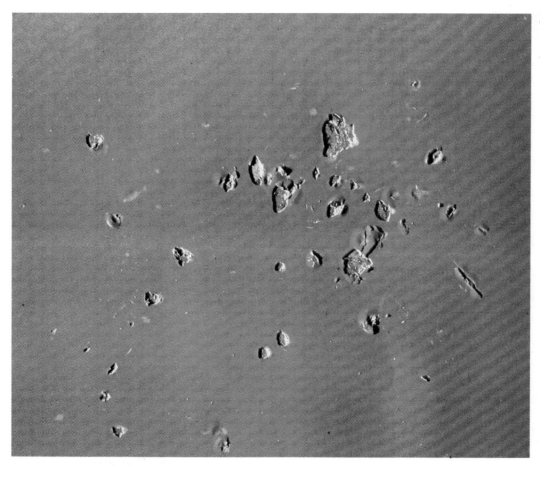

Plate 25   Stone chipping

Stone chips are small areas of damage to the paintwork caused by stones or loose chippings.

Causes

Stones or other hard substances (e.g. loose chippings) hit the vehicle with varying amounts of force (speed). Depending on the force of the impact not only the top-coat but the entire paint film build can be destroyed. Moisture can then penetrate the areas causing corrosion and further paint detachment.

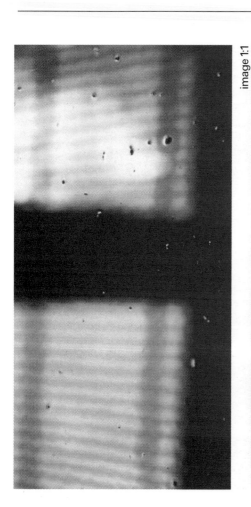

image 1:1

80 x magnification

Plate 24   Dust contamination (dirt, bits)

This contamination is due to visible dust particles of various sizes and forms which are embedded or form raised spots in the topcoat.

Causes

Dust contamination can be caused by various problems:

1. Inadequate cleaning of the vehicle after sanding.
2. Unsuitable working clothes which leave dust and dirt fibres.
3. Dust problems caused in the booth, e.g. the dust filter is dirty or leaks, the pressure bal-ance is incorrect, the booth is very dirty or there is no or inadequate filtration of the pres-surised air supply.

image 1:1

Plate 27   Rust (corrosion)

Subsurface rusting is visible as paint damage due to irregular bumps in the paint (blisters). If the blisters burst or crack, brown spots of corrosion (rust) are visible. In the case of an aluminium body, white spots of corrosion (white rust) are visible.

Causes

1. Moisture penetration, due to cracks or mechanical damage (e.g. stone chips) to the paintwork, right down to the bare metal.
2. Before finishing rust was not thoroughly removed.
3. The surface of the metal has been contaminated, e.g. with hygroscopic salts or sweat from hands. Rust and water blisters form due to the reaction with moisture from the air.

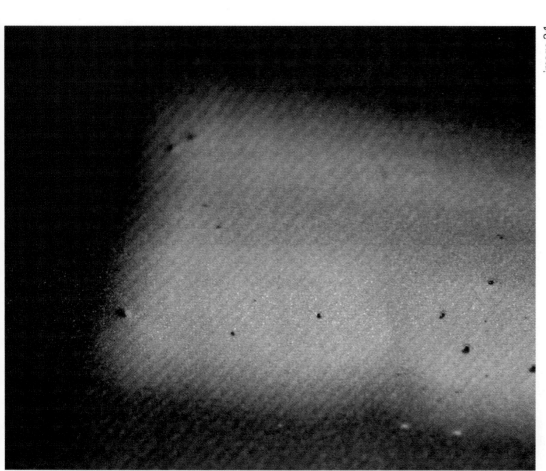

- image 2:1

Plate 26   Seed (specks)

Seed develops in varying shapes, sizes and number which are embedded in the surface of the paintwork.

Causes

1. The use of paint which has been kept longer than its recommended storage life.
2. The addition of incorrect hardener or thinner.
3. The use of re-thinned 2K materials whose potlife has already been exceeded.
4. Pigment conglomeration due to insufficiently stirred material.

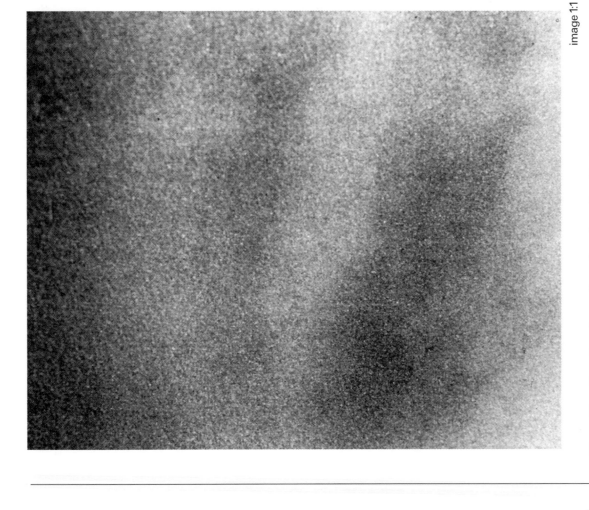

image 1:1

Plate 29 Clouding (mottling, floating, flooding)

Clouding is a speckled appearance of metallic finish which covers a large area.

Causes

1. The basecoat has been applied in uneven coats.
2. The flash off of the basecoat was too short before the clearcoat was applied. This can cause the metallic coat to be redissolved. Result: the metallic and pigment particles float and change their position.
3. Too wet a spray application to the first clearcoat, causing a redissolving of the basecoat.

image 4:1

Plate 28 Water spotting (rain spotting, water marking)

Water spotting appears as circular, mainly whitish, spots/marks on the surface of the paint.

Causes

Water spots/marks appear if water droplets (rain or dew), together with pollution from the air (e.g. dust, chalk or salt), dries on to the surface. Normally no damage appears within the circular marks, only the edges are seen as raised marks. The problem of water spotting only occurs with freshly painted paint finishes which have not been thoroughly dried/cured.

Chapter 6 | Surface Preparation and Priming

Surface Preparation

Health and safety at work and PPE

Dry sanding is the favourite method of flatting as it is fast and a good finish is easily obtained. There need be no danger to health, providing always that the correct PPE is worn and the waste is handled properly. This can be seen in any bodyshop where adequate dust extraction is installed and the personnel adopt the proper procedures. The atmosphere is free of dust, there is a minimum of dirt and dust on the floors and the painter has no difficulty in keeping the spray-booth clean. Moreover, every job can come out of the booth in finished condition and, with the exception of the occasional minor blemish, be ready for fitting-up. We will look at dust extraction in more detail later in this chapter but first we examine protecting your health and safety with PPE, together with some other more general topics.

The dirt and dust made at this stage of repair can be a hazard to health for everyone who works in the bodyshop, even to the extent that it could eventually prove fatal. The first part of Chapter 4 looked at why such dust can be so damaging and what the end result could be. If you have not read it, do so now.

Housekeeping and waste disposal

It is more comfortable for you and more profitable for the business to keep the workplace clean and tidy. You know, too, that it is safer, quite apart from being a legal obligation. This is especially important where waste from filling and finishing is concerned. Much of the material that is used carries warnings about self-ignition (spontaneous combustion) in certain circumstances. This means that dust, wipes, mixing palettes, emptied containers and solid waste must be put into water wetted waste bins with lids and dampened regularly. This is classified as Controlled Waste and must be collected by a registered contractor.

Here are some important points to keep in mind when disposing of your waste:

- The peroxides used in fillers may interact explosively with cobalt, which is present in some accelerators.
- Peroxide hardener can ignite if it comes into contact with air-drying paint.

Hazard	Protection
Dust and particles from sanding panels and filler	Nuisance dust masks to BS6016 type 1 or 2, HSE approved or to EN 149 class FFP 2S EN 149 class FFP 1 may be acceptable for short periods of local repair work
Fumes and mist from solvents	Activated charcoal mask to BS 6016 type 1 for brief exposure or EN 149 FFP1 Activated charcoal filter to BS 2091 for prolonged exposure (except isocyanate) An air-fed mask must be used for isocyanates
Pressurised gases such as compressed air can be extremely dangerous if used in an uncontrolled manner	Compressed air and gases must only be used with equipment that provides full shut-off and directional control. Normal workwear provides no protection against pressurised gases
Mains electricity is as dangerous as pressurised gases if used carelessly or the equipment is poorly maintained	Plugs, cables and equipment must be in good order without defects Cables and plugs must be protected from cuts, chaffing, crushing or any form of tension
Other people in the workplace will also be affected by dust and fumes The accumulation of dust also has an adverse effect upon the quality of paintwork	Dust and fume extraction, preferably by a centralised system, to reduce levels well below the Occupational Exposure Limits General extraction in no way replaces PPE

- For a similar reason peroxide hardeners must not come into contact with dry, nitro-cellulose sanding dust.
- Some epoxy hardeners contain amines (an ammonium compound), which can also interact with nitro-cellulose sanding dust.
- The waste bins should be kept away from any source of sparks or naked flame

Good housekeeping is not only being safe; it is, in many ways, the key to successful refinishing. It follows quite naturally from thinking the job through and planning each process. Waste materials and items no longer needed are removed to the appropriate storage. Equipment is cleaned immediately and is racked ready for the next use. Maintenance and general cleaning is dealt with in good time and whenever necessary to cater for the flow of the work.

Cleaning and tidying is part of every job, as well as carrying out your Duty of Care under the Waste Disposal Code of Practice.

Sources of information

Trade journals are a useful source of information on various aspects of refinishing. Articles and booklets are frequently published giving general guidance on the topic and up-to-date information on materials and processes.

Equipment and tooling

The initial examination of the finish will need wipes, solvents and a sharp knife or blade. There are two pieces of equipment essential for paintwork examination. The first is a paint depth gauge which may be mechanical or electronic. Mechanical gauges work on the principle of magnetic attraction. Because of this they are unsuitable for aluminium or GRP bodies. The electronic type is perhaps the best as they can be easily calibrated and are available in versions that can read paint depth on any metal substrate. The other indispensable item is a magnifying glass, or an illuminated viewer.

Examination of the surface must also involve colour. If an area well lit by daylight with a neutral coloured background is not available, simulated daylight lighting should be provided outside the paint booth for colour comparisons. Where the paint maker specifies a particular light source, this must be used. In some bodyshops you may have the use of a colour comparitor or other electronic instrument to assess colour.

Many bodyshops continue to use disposable masking or vehicle covers but there are also reusable types which may prove cost effective. Wheel covers are often of the substantial and reusable type.

Sanding is best done using a random orbital sander for normal work and a half sheet flat bed sander for larger areas. Pneumatic units of the dust extracted type to link into the extraction system are preferred. The latest materials demand units with an orbit of no more than 6 mm. A low orbit should be combined with the use of a firm but flexible backing pad. A selection of hand sanding blocks and files are essential. These too should be of extracted type wherever possible (see Fig. 98).

Wet flatting will call for suitable blocks, buckets, sponges and leathers. Large amounts of such work are best done on a wet deck, a floor with a grid which provides drainage. Every bodyshop will need to do some wet flatting occasionally and so a sink and clean water supply is essential.

Job	Source of information
PPE selection	PPE makers' guidance notes and sales brochures Specialist suppliers' information Product makers' data sheets Vehicle makers' information MIRRC (Thatcham) information
Substrate recognition	Vehicle makers' manuals, training material and sales brochures Paint makers' information MIRRC information
Surface preparation	Paint makers' manuals Abrasives suppliers' information Equipment makers' information Factors' information sheets Vehicle makers' manuals and service letters
Use of chemical preparation materials	Makers' directions, data sheets Paint makers' manuals Vehicle makers' manuals, service letters and training information
Masking	Product makers' notes General press articles Training information

Figure 98 An air-extracted hand sanding block

Masking plays a major part in successful refinishing. A fully equipped trolley to dispense plastic film or masking paper, together with automatically applied tape, is another essential.

Preparation and painting is often called for on single panels, prior to fitting them to the vehicle. A suitable stand is an ideal way of supporting panels (see Fig. 99). The best commercially made stands can hold panels ranging from small wings up to large van doors.

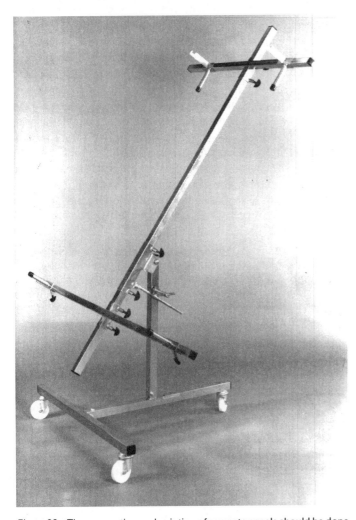

Figure 99 The preparation and painting of separate panels should be done on a suitable stand

Recognising substrates

The substrate is the surface to which the finish will be applied. This can range from bare metal through to previous paint films. Fillers and paint coatings have some versatility regarding the surfaces to which they will adhere. However, to avoid poor adhesion and other problems it is as well to know as much as possible about the surface to be painted.

The bare or pre-treated surfaces that you will be called upon to finish are generally easy to identify. The dark grey of steel, the silvery sheen of galvanisation and the dull, almost creamy texture of aluminium do not generally cause a recognition problem. Glass reinforced plastic (glass fibre) may be any colour although the thickness gives it a dull sound and heavy texture. Plastic is notable for flexibility and is usually now marked on the reverse side with the material code.

Where some difficulty may be experienced is in determining whether the coating on a new panel is phosphate, electrophoretic primer or just transport protection. There have been many instances where a permanent anti-corrosion treatment has been removed in the belief that it was temporary protection. If this was done on part of a vehicle still covered by an anti-corrosion warranty, any subsequent corrosion of the repaired area would be the responsibility of the repairer. To overcome such a risk the coating must be identified. If in doubt ask the supplier of the part.

Most refinishing involves filling, priming and painting onto existing paint coatings, except for body shell changes for some makes. So it is also important to know something about this type of substrate.

There are three main aspects to consider:

■ the type of finish material already on the surface,
■ the condition of that finish and the substrate below, and
■ the depth of the coatings.

The material on the vehicle will determine which type of finish or sealing coat should be applied if complete stripping is not an option. This can be carried out by a solvent test. A small piece of wipe is soaked in cellulose thinner and wiped across a part of the body that will be sanded down:

■ Immediate dissolving indicates a cellulose finish.
■ Slow dissolving after a short, covered soak indicates that it is probably a thermoplastic acrylic.
■ No reaction means that the paint is a two-pack acrylic or equivalent.

Thermoplastic acrylics must be treated particularly carefully because the finish can reflow when heated, as well as being affected by some solvents.

It may also be necessary to check for the presence of a clear lacquer coat, particularly on metallic finishes. Dark, solid colours are also often lacquered to preserve their appearance and add lustre. There are two tests that may be used. Taking some shavings from the surface with a very sharp blade is one. If the shavings are white, it is lacquered. Alternatively, rub the surface with a burnishing compound. Colour will be removed if it is solid colour or unlacquered basecoat. The cloth will not discolour if it is lacquered.

The depth of paint film will also determine how the repair is undertaken. Most paint makers are unhappy with finishes that exceed between 250 microns and 350 microns in total depth. Most car factories produce finishes ranging between 80 and 120 microns. A typical repair process is likely to add between 40 and 80 microns. If the panel has already been repainted, the paint depth could already be in the region of 200 microns. It may be considered worthwhile in such a case to reduce all the paint to a depth suitable for the new coatings (see Fig. 100).

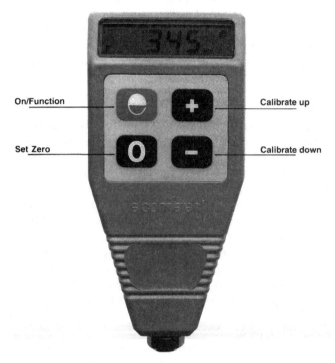

On/Function
Set Zero
Calibrate up
Calibrate down

Figure 100   Finish depth must be checked with a suitable gauge to ensure that permissible limits are not exceeded by the refinishing operation

Here is the calculation for a paint depth not greater than 250 microns:

	microns
Original factory finish	120
Rectification or repair	80
Total depth of paint film (on the vehicle)	200

Paint preparation that will be needed:

Paint on vehicle now	200
Proposed repaint	80
Total possible	280
Permitted depth	250
Excess to be sanded off before painting (minimum)	30

There is one other point to bear in mind when looking at paint depth. There is a tendency for the depth of paint to be greater on the horizontal panels, such as the roof and bonnet, than on the vertical surfaces. This seems to be true of both repaired and factory finished vehicles, even though many manufacturers use an electrostatic process, which should provide uniform thickness over the entire vehicle.

The condition of the finish and the substrate is the basis for obtaining good adhesion. Any bubbled paint is an almost certain indicator of surface corrosion or contamination on the metal itself. There is no alternative to stripping back to bare metal in this area at least. Light corrosion may be dealt with by using the derusting cleaner supplied by the paint maker. Serious corrosion must be neutralised chemically or removed by shot blasting.

Finish condition itself can be divided into a number of categories:

- hardness or softness,
- adhesion problems,
- film failure or contamination.

When the paint film is being cured, particularly where factory finishes are concerned, the temperature of the vehicle body may become either too high or not high enough, depending upon the time spent curing in the booth. The former results in paintwork that is too hard and in the case of the latter, too soft. Such finishes should be removed to the layer below or to the next layer that is in sound condition and has been correctly cured. Correctly mixed but soft paint may be recured.

The second aspect of condition, adhesion, is revealed by excessive stone chipping or greater flaking at each point of impact. Stone chipping is a common form of damage, particularly amongst those to whom the exhaust pipe of the car in front exerts a strong power of attraction! It is the nature of the stone chips that interests us and what is revealed at the bottom of them. There are two reasons for poor adhesion: poor preparation and painting, or overbaking. Typically with this defect, the paint easily parts company from the primer filler coat (undercoat). This is all too easily seen because the primer filler coat is either a different colour or a matt version of the same colour. The crater will appear to be steep sided with a flat bottom when viewed under a magnifying glass. In extreme cases, large, circular discs of paint will have come away. Applying a piece of masking tape to damaged paintwork can be helpful. Loose flakes pulled off by the tape are a certain indicator of poor adhesion.

The definitive way of checking adhesion is by carrying out a 'cross-hatch test' with a multi-toothed cutter. The tool is dragged across the paint in two directions, making cuts in the shape of a grid. The degree of breaking away at corners or of whole squares of paint gives an indication of the quality of adhesion. The test is obviously destructive; the panel must be repainted whatever the result. Some vehicle manufacturers use a test that has been specifically devised for this problem. This consists of a pneumatic gun which blasts shot at a small square of paint. Damage is compared to a chart to determine defective paintwork.

If poor adhesion is caused by an overbaked topcoat, only the topcoat need be removed. If the poor adhesion is caused by an overbaked primer filler coat, the filler itself will appear smooth and glossy and must also be removed. Where the defect is caused by poor painting procedure, the defective coats must be removed and a sound layer prepared correctly before refinishing can begin. Overbaking is usually a fault in original production; poor painting almost always a refinishing defect.

The third condition likely to cause concern is the state of the paint film itself. Vehicles untouched from production are unlikely to have any problems other than hardness, softness or colour fading. Those that have been repaired may have almost any combination of materials, sometimes in a poor state. Suspect and obviously unsatisfactory finishes must be removed down to a good substrate or in their entirety if there is any doubt. Sealing coats may only be used where there is incompatibility, not coating failure.

Chemical preparation

Vehicles collect a variety of contaminants on their surfaces during everyday use, mostly described as 'traffic film'. To these must be added preservatives in the form of waxes and polishes which may be applied to the paint film, and traces of lubricants thrown out by vehicles. As a painter you will be only too well aware that the worst of these contaminants is silicon. To avoid problems of contamination the surface must be cleaned very thoroughly before starting preparation. This should be done in two stages. First a wash, preferably with a high pressure washer, to remove ordinary soiling and road dirt. This is followed by wiping over the areas for repair with the general surface cleaning solvent from the paint system. These chemical cleaners are designed to remove wax and other stubborn coatings.

The cleaning treatment should have already been carried out on body repair jobs. If the work is refinishing only, then the initial wash to remove ordinary dirt and contaminants should be carried out before the vehicle enters the paintshop. That should be followed by chemical cleaning as soon as possible.

At every stage of the refinishing from now on, the chemical cleaner or another designated wiping solvent must be used to keep the surfaces free of contamination. It only takes a touch from an apparently clean finger to create a blemish on the finish.

It may also be necessary to strip paint from panels chemically using a paint stripper. This is sometimes a specific requirement, particularly where paint must be removed from galvanised body parts. The use of paint stripper avoids any damage to the galvanised coating. It is also sometimes more convenient to strip defective paint coatings with paint stripper. What may be called heavy duty paint strippers can be extremely dangerous and irreversible skin damage can result from contact with them. It is imperative then that full protection be worn when handling these materials. Be guided by the information on the data sheet which must, by law, accompany them. In general, however, full protection against splash must be worn and a respirator if this is indicated.

It also follows that the cleaning must be done in an area where other personnel cannot be affected. Waste disposal must be carried out correctly.

Vehicle protection

Protection and masking do have much in common and both are aimed at protecting the vehicle. However, attaching masking materials has as much to do with the techniques and the results of painting than with protection, so I have chosen to deal with it separately.

When vehicle bodies are repaired, the effort is concentrated on restoring the shape and finish of the body. Very often, little thought is given to the mechanical and electrical systems, or even the interior. The vehicle is constantly moved from one area to another during the refinishing process and seat covers become soiled or even damaged through workers getting in and out of the driving seat wearing protective clothing. This is understandable, but can be expensive.

The wheels, running gear and underbody are all areas out of the line of sight for much of the time and so tend to be neglected. A wheel cover need only be forgotten once and the telltale signs are all too evident against the murky black of the brake, the inner wheel surface and the tyre itself. Some refinishing materials may also be damaging to the materials from which the tyre is made. These critical components should be protected at all times during the repair process.

The interior of vehicles is even more costly to valet and return to original condition. Plenty of covers carefully fitted will more than repay the time and trouble of putting them in place. All this protection must be constantly checked to prevent openings from occurring.

Where the refinishing is part of a repair, components will have already been removed and stored safely. If the job only involves refinishing, additional components may need to be removed before work can begin. The procedures outlined in Chapter 3 under Dismantling for Repair should be followed.

Some components may also need to be considered for special protection from the effects of heat during baking. Booth temperatures will not normally be of concern. Infrared curing may pose a problem where it is possible for a panel temperature to rise too high. A check should be made for electrical control boxes or relays which may be attached to the inside of panels. Side panels of the passenger cell and the luggage boot are probably the most likely locations for these additional components. These units must either be removed or the panel temperature carefully monitored during the curing period. Panel temperatures should not greatly exceed 80° Celsius, although a brief rise is unlikely to cause damage to other components. Removal procedures are outlined in Chapter 3 under Dismantling for Repair.

Masking

Masking is an art that can make or mar paintwork. You may consider this an overstatement, but there can be no doubt that careless or inexpert masking is all too easily seen in the end result. Pronounced edges, noticeable colour differences and overspray all reveal lack of care or skill. On the other hand, discretely hidden edges, smooth blends and no signs of overspray create confidence in the customer and add to your pride in a job well done.

Masking tape will be needed somewhere, regardless of the type of masking in use. Strong, good quality masking tapes, with a noticeable crepe effect, are preferred. Several sizes ranging from narrow to wide will help in achieving the correct result.

There are a number of options for the bulk covering. The main part of the vehicle may be covered with a reusable cover, and disposable paper or plastic film used adjacent to the repair. Alternatively, reusable film can be used for the whole body. Spray-on masking is yet another option. Whichever type is used, it is important to cover the vehicle neatly, to avoid dust holding crevices and openings into which spray may enter. Special materials can be used to save time. A 'roll edge' can be purchased ready made. There is packing strip available to lift window rubbers and other soft trim (see Fig. 101). Alternatively, a suitable size of flexible wiring or tube may be used.

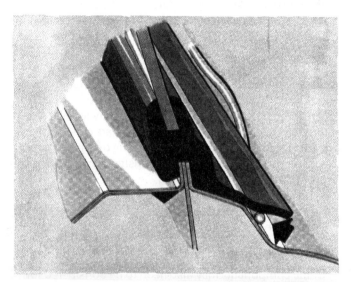

Figure 101 Weatherseals should be lifted by cord to provide fade-out and prevent unsightly edges

The principle behind all masking is to protect those parts which should be kept clean and to help in achieving a good result on those that are painted. Sharp edges should be avoided wherever possible. Door pillars, for example, can be blended-in by using the 'roll-edge', or better still a hooded cover. The 'roll-edge' is ideal for local repairs on large panels such as bonnets or for fading-out at the edges of full panels (see Fig. 102). The movement of the spraygun over the rounded edge provides a degree of fade-out. Where sharp edges are inevitable, arrange for the edge to be around the turn of the panel or hidden behind trim. So, for example, a front wing could be finished in the bonnet shut.

Gaps must be closed to prevent paint entering or dust escaping. Where the panel needs to open, a flap can be made to seal on closure by backing up the protruding edge with another strip of masking (see Fig. 103). Masking under the sills and ends of the body must be sufficiently long to wrap underneath and be held up from the floor. When wheel covers are used, do consider the space behind them. The wheel arch and suspension should be protected from paint and dirt, particularly the pivot rubbers, gaiters and exposed sliding members of struts and shock absorbers.

If you use compressed air to release dust when tacking off, the masking must be able to withstand the air jet. Openings or unsecured edges could easily rip. Remember, too, that condensation can prevent adhesion of the tape, particularly

to surfaces such as hard trim and glass. If the bodywork is just a few degrees below the ambient temperature, a fine moist film will cover these colder surfaces. A convenient way of warming up the repair area is to use infra-red heaters for a few moments; hand held for a small repair or a stand unit for larger jobs.

Here is your checklist for masking:

- Use good quality masking tape.
- Use masking paper, film or spray made for the job.
- Keep the covering neat to avoid dust traps.
- Close up all openings.
- Make stiff flaps for panels that must open.
- Use the 'roll-edge' or a hooded cover for all blend-in areas.
- Hide all sharp edges in shuts or under trim.
- Packing can hold rubbers away from the body.
- Protect hidden spaces such as inside wheel arches.
- Wrap under sills, and at the front and rear.
- Dry off condensation.

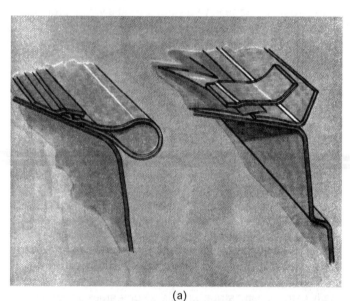

(a)

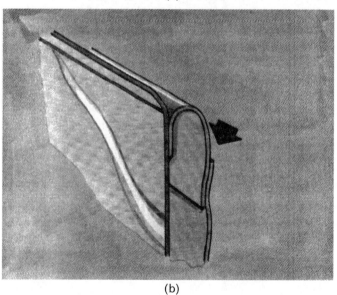

(b)

Figure 102 (a) The gradually reducing depth between the roll-edge of masking and the panel helps in avoiding edges by giving some fade-out, (b) pull back the roll before curing to prevent a hard edge forming

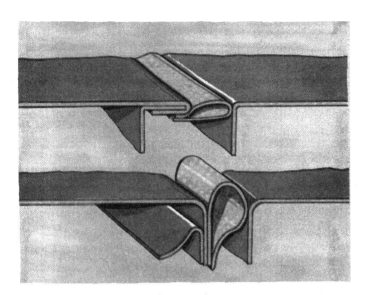

Figure 103 Rolled masking bridges gaps effectively and provides fade-out and an absence of hard edges if it is pulled back before curing

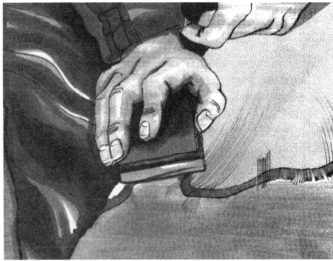

Figure 104 For wet-flatting large areas use a rubber block; note that gloves have been omitted for clarity

Wet and dry sanding: dust extraction

In these days of tough bargaining with insurance companies, inability to produce 'from the gun' paint finishes can mean financial loss, instead of modest profit. Tight estimating does not allow the polishing of every panel; it should only be necessary to remove the occasional blemish from new paint. To achieve such standards of finish, a high level of cleanliness in the booth and throughout the bodyshop is essential.

The same cost constraints demand a fast rate of sanding, while the customer expects good quality. For all these reasons most bodyshops now use dry sanding, coupled with effective dust extraction, for all normal work. This section then is mainly concerned with dry sanding and it is assumed that it will be carried out using extraction, preferably a centralised system. Where extensive wet flatting is still carried out this should be done on a wet deck to overcome the dust and wet mess inevitable with normal flooring (see Figs 104 and 105).

Any components on, or next to, the work area must be removed and the vehicle masked up. There may also be a need to mask up behind apertures to prevent soiling behind the opening. Be especially aware of electrical cables and exposed terminals. Wrap them in a piece of masking to prevent soiling.

Before commencing sanding check that the extraction system is in working order and that the PPE you need is to hand. A supply of suitable abrasives must be obtained with the correct extraction openings. These openings vary depending upon the equipment in use. Random orbital sanders have holes in the centre or on the outer periphery (see Fig. 106). Half sheet sanders have a row of holes each side but the spacing will vary again depending upon the make of the machine. The fastening will be either adhesive or the fabric hook type. The grades of abrasive are selected for speedy removal of material and then a gradual change to finer finishing and feathering of the surrounding edges of the existing finish.

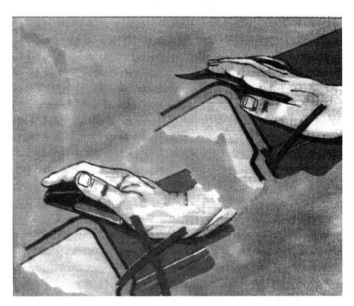

Figure 105 When wet-flatting edges and sharp corners use gentle pressure and the flat of the hand

Figure 106 Correct alignment of the extraction holes on both sanders and hand sanding blocks is essential for efficient dust pick-up

Details of the grading system and general guidance on initial sanding will be found in Chapter 4. The flatting of foundation coats, usually primer fillers, is generally done by dry sanding although there are some products where wet flatting is recommended. A guidecoat is often applied wet-on-wet at the end of primer filler application. If this was not done then a light dusting coat of a contrasting colour should be given to the surface. This is of great help in sanding and maintaining a smooth contour.

A random orbital sander fitted with, typically, P400 abrasive is used to flat the whole area and provide an essential key onto the existing paintwork. The abrasive range is quite wide and paint makers suggest grades from as low as P280 through to P500. It may also be considered desirable to finish one or two places with a hand block and a suitable paper. Never be tempted to use a machine disc by hand. A P400 disc creates scratches 3.6 microns deep when used on a sander. When used by hand the grooves will be around 7 microns deep. All dry sanding should be extracted, of course, and the operator should wear a suitable dust mask as an additional precaution.

Dry sanding abrasives are lubricated with stearate powder, which must be completely removed before any painting is done. This is easily accomplished using a chemical cleanser and degreaser from the paint maker's range of products. The area must be thoroughly wiped, including any recesses.

Wet flatting of these surfaces would call for the use of paper grades in the region of P600 or P800 (see Fig. 107). Be very careful to check the paint maker's data sheet when using an unfamiliar product. Some may not be wet flatted. Always use a clean bucket, sponge and water to avoid contamination. Always wipe off the residue of flatting before it dries with a leather. The leather is a likely source of contamination and should always be cleaned, squeezed (not wrung) and stretched flat to dry between use.

The final stage of preparing foundation coats is to use an abrasive pad (see Figs 108 and 109) and a flatting paste to provide a keyed surface for paint adhesion that is also chemically clean. It must be wiped again before painting to remove any last traces and particularly finger marks (see Fig. 110).

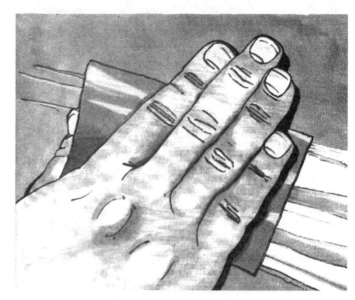

Figure 108 How the abrasive should be held during hand flatting; gloves have been omitted for clarity

Figure 107 Fine wet-flatting requires the use of a soft block

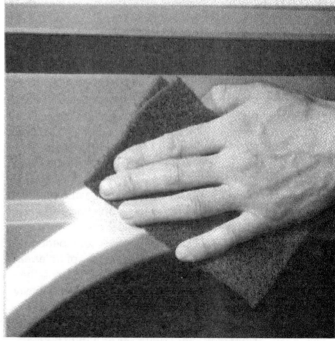

Figure 109 A woven abrasive is used to flat glossy finishes

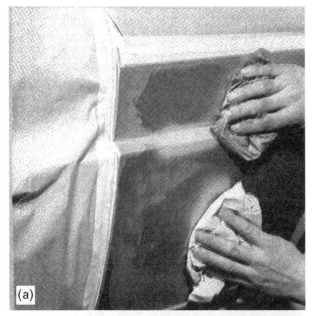

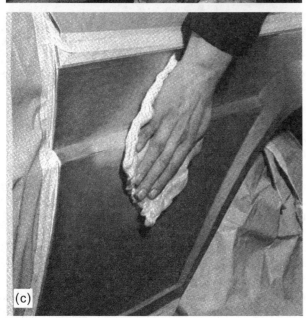

Figure 110 The routine sequence of (a) spirit wipe, (b) blow-off and (c) tacking

Here are the reminders about sanding and a few pointers on wet flatting from Chapter 4:

■ When about to dry sand ensure that the extraction system is working correctly.
■ Have enough suitable dust masks handy to protect you through the session.
■ Wear gloves.
■ Use new discs to break the surface. Renew discs as soon as the cutting action falls away.
■ Ensure that extraction holes line-up correctly.
■ Place the sander in position before switching it on.
■ Keep the speed low.
■ Let the weight of the machine provide pressure on horizontal surfaces; use similar pressure for vertical panels.
■ Keep it moving to avoid heat build-up and work progressively over the area.
■ Stay inside the final repair area until you are ready to feather the outside edges.
■ Work away from sharp edges or curves unless stripping these areas.
■ Every effort should be made to preserve surface galvanising and maintain the panel thickness. Never use a grade coarser than P180.
■ Change up to finer abrasives in stages no more than two grades apart.
■ When using wet flatting for preparation use clean water to avoid contamination.
■ Change the water for every job.
■ Rinse and leather off the residue of wet flatting before it dries.

Fillers and Foundation Materials

Health and safety at work and PPE

Many of the hazards associated with refinishing affect us in a variety of ways. In this section those which can harm the health of the painter and others nearby are discussed and the preventive action is outlined. The next section, Housekeeping and Waste Disposal, deals with such matters as the proper handling of dangerous waste. The later section on Material and Equipment Storage also deals with safety and concentrates on how materials should be stored and used.

The main hazards to personal health in this work are solvents and paint spray. The two factors may be combined so that protection should prevent the inhalation of spray dust particles and the fumes given off by solvents. Up to 70% of paint sprayed onto a surface can be lost in conventional spraying. However good the booth extraction may be, there is a high risk of lung damage from the paint particles. The amount of free spray in the air is much reduced where high volume, low pressure (HVLP) sprayguns are used. Some makers claim a reduction of waste spray to as little as 11%. Nonetheless, continuous exposure to low levels such as this would still pose a high risk of lung damage.

Where solvent based materials are still used, the PPE should be capable of preventing any exposure. The isocyanates present in the hardeners of most 2–pack paints are familiar to all painters and can easily cause sensitisation. That is a condition where the tiniest trace of the irritant can bring on a severe asthmatic attack. There are reported instances where workers outside the spraybooth have been sensitised in this way by inhaling a tiny trace of leaked fume over a number of years. That is why paint booths must have no more than equal or negative pressure compared to the outside working area.

Full, air-fed face masks are the ideal way of preventing such contamination, provided that clean, cool air is supplied to them. Brief exposure protection may be obtained from cartridge type respirators although these do have the risk that once the cartridge is used up, protection from fume is no longer available. Only air-fed face masks may be used when spraying materials containing isocyanates.

Protection from dust has already been dealt with at some length, and you can find this information in earlier chapters.

Suitable gloves should be worn to protect your hands. The author knows from personal experience that constant contact with solvents can damage the skin and lead to problems in later life. The overalls worn during work outside the paint booth may be of the conventional industrial type. Whenever final preparation is in hand, and particularly when working in the booth, painters' disposable overalls should be worn. These have no loose fibres and minimise the chance of dust, dirt or other inclusions in the wet paint film. The majority of fibrous inclusions come from the clothing of the painter or others who have been near the vehicle or panel. One important reminder! Always bear in mind the danger that can arise from using compressed air. In particular, painters often blow off their overalls with the air line, an action which can be very dangerous. Wear the correct overalls and never use compressed air at high pressure to blow them off.

Housekeeping and waste disposal

All waste from the painting operation is potentially dangerous. Paint materials themselves can be very toxic due to isocyanates or other ingredients. Many paint containers and data sheets carry warnings about self-ignition in certain circumstances. Masking and protective sheeting is often potentially flammable, quite apart from the paint adhering to its surface. It makes sense, then, to comply with the various regulations which control the disposal of waste and in so doing look after yourself.

Liquid waste must be emptied into a drum provided for that purpose. If the paintshop in which you work has a thinners recovery unit then there may be more than one container (see Fig. 111). Dry or cured waste which involves materials that may ignite spontaneously should be put into flammable waste bins. These containers usually have a small quantity of water in the bottom to help prevent self-ignition. Your bodyshop will have a contract with a registered contractor to remove dangerous waste under controlled conditions. In the meantime it must be stored securely against the elements and protected from interference by vandals. While the waste is on the premises, and in transit, it is the respon-

sibility of the bodyshop. Here again are the important reminders about handling dangerous waste:

- The peroxides used in fillers may interact explosively with cobalt, which is present in some accelerators.
- Peroxide hardener can ignite if it comes into contact with air drying paint.
- For a similar reason peroxide hardeners must not come into contact with dry, nitro-cellulose sanding dust.
- Some epoxy hardeners contain amines, which can also interact with nitro-cellulose sanding dust.
- The waste bins should be kept away from any source of sparks or bare flame.
- Empty cans filled with fumes are more dangerous than full ones; they must be removed immediately.

Figure 111 A solvent distiller for recovering solvents in the bodyshop for re-use

Cleaning up dry paint residues should be done regularly, perhaps every week. The scrapers and brushes used should be of non-ferrous metal or a stiff plastic to avoid sparks. The swept rubbish should be put into a metal container, damped down, covered and stored outside the building. Spillages that occur during work should be absorbed into sand or a special inert material and disposed of in the same way.

Spraygun cleaning is an activity which is specifically controlled by regulations brought in under the Environmental Protection Act (known as the EPA). An automatic, totally enclosed machine must be used (see Fig. 112). Testing and spray-out should also be done into the same or a comparable machine with extraction running. Paint or solvent sprayed from the gun must be enclosed and captured.

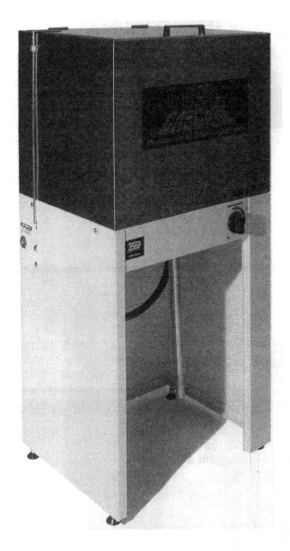

Figure 112 An automatic spraygun cleaner suitable for small bodyshops

Sources of information

As a painter you have the advantage of having one of the best information systems of any trade. Here are the topics that you may need and the sources from which the information may be obtained:

In addition to the usual written and microfiche information, the makers of paint products generally offer a very good telephone advisory service, backed up by area representatives who can advise and give hands-on demonstrations. In cases of unusual difficulty, they have the backing of full analytical laboratory facilities. The makers of booths and infra-red curing equipment also provide information and assistance in answer to queries or to resolve problems. Vehicle makers may also be able to give specific advice relating to difficulties with unusual finishes or substrates.

Many of these companies, and organisations such as the MIRRC and specialist training bodies, can provide a full range of training courses. The training centres of the major paint makers are also approved for NVQ assessment.

Equipment and tooling

The painter needs some of the most expensive equipment in the bodyshop, if high standards of finish and the requirements of regulations are to be met. The paint booth is the most commonplace of these. This can be supplemented by infra-red equipment which is available in many versions ranging from small hand-held units through many sizes and types up to a fully automated, computer controlled arch. To handle a high throughput an infra-red drying arch may work alternately between two booths.

Paint storage and mixing is another activity which must be controlled and is also governed by regulations. A mixing room, and perhaps a separate paint store, is a necessity. The facilities here will include a paint mixing scheme comprising a motorised rack in which ready to use cans of base colour are stored. Very accurate scales, usually electronic, are required together with a microfiche reader and storage facility. Some manufacturers provide computer based formulations linked to the electronic scales. There may be other items relating to the particular make of paint in use.

Smaller equipment will include a gun washer for cleaning sprayguns after use. This should be a fully enclosed, automatic unit which prevents the escape of solvent and fume to comply with the regulations made under the EPA.

A variety of sprayguns is needed, of course, dictated by the type of work undertaken, the materials in use and, per-

Job	Source of information
Cleaning surfaces	Product information sheets Paint makers' data sheets and manuals Vehicle makers' information Suppliers' literature Training information
Flatting and preparation	Paint makers' data sheets and manuals Product makers' information
Mixing	Paint makers' data sheets and manuals Suppliers' information
Application and curing	Paint makers' data sheets and manuals Equipment makers' information

haps, the preferences of the painter. Most importantly, the EPA dictates which type of equipment may be used for some operations. An HVLP spraygun must be used for the application of primers, primer fillers and surfacers (see Fig. 113). A variety of fluid tip and needle set-ups must also be available, together with spraygun service kits and cleaning tools. One important point to note is that the fluid tip and needle are often made as a matched pair and should only be interchanged when they can be obtained separately (see Fig. 114).

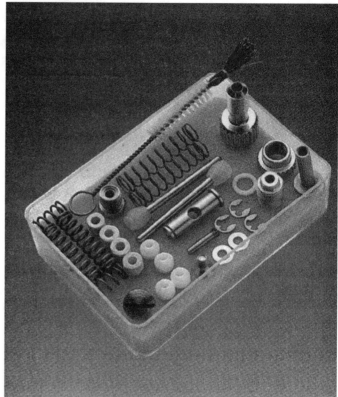

(a)

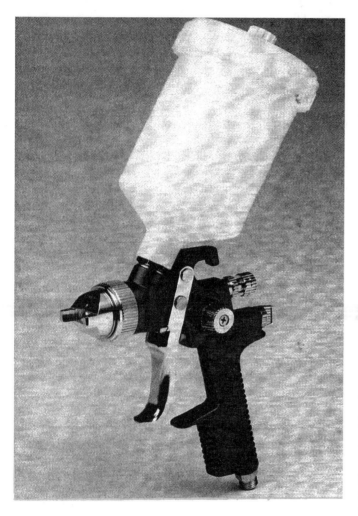

Figure 113   An HVLP gravity-fed spraygun

Two small but essential tools upon which success will depend are an accurate seconds timer and some viscosity cups. Mixing 'sticks' and cups are equally important. Stainless steel mixing and measuring jugs are also most useful. For mixing primer fillers, which are not normally kept on the mixing scheme, a paddle fitted to a pneumatic or cordless drill will prove invaluable.

The successful application of primer fillers and topcoats is dependent upon having sufficient high-quality air. The compressed air plant itself must be capable of delivering sufficient cool, clean air to provide the spraygun and the air-fed mask worn by the painter with all that is needed (see Fig. 115). A continuous supply of at least 0.56m³ (20 cubic feet) of air per minute is needed for sprayguns currently in use.

(b)

Figure 114   (a) Spares should be on-hand for all your sprayguns; (b) the needle-set is made as a matched assembly, but the air-cap may sometimes be interchanged if it is permitted by the spraygun maker

The only satisfactory way in which the adequacy of the supply may be monitored is to equip the spraygun with an air micrometer (see Fig. 116). This is a combined pressure gauge and adjusting device which is attached to the air inlet of the gun and enables the air supply to be continuously checked and adjusted if necessary (see Fig. 117). The pressure gauges on the regulator or water separator/filter unit do not indicate the pressure available at the spraygun. HVLP guns should also be checked for pressure at the air cap occasionally. A special air cap fitted with a gauge is available which shows the operating pressure (see Figs 118 and 119).

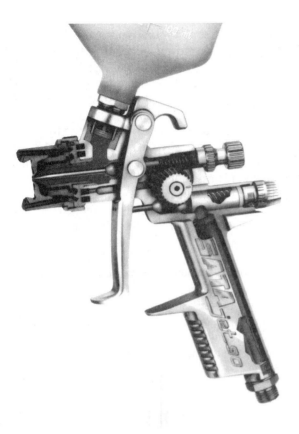

Figure 115   A sectioned view showing the paint and air passages inside a spraygun

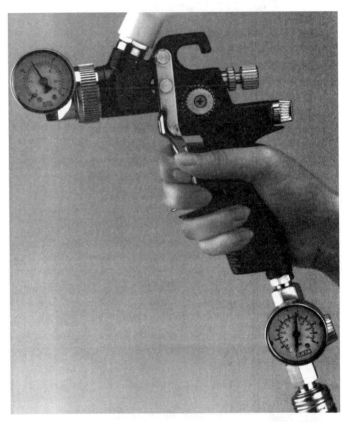

Figure 117   HVLP sprayguns must not emit more than 12lbf/in² at the air-cap; on some makes of spraygun one gauge will suffice to check this

Many painting problems can be put down to insufficient air volume, which results in falling pressure. It is of little value to set a pressure on the filter regulator if there is not enough air to meet the demand. If there is not enough air, the pressure will drop. The continuous pressure available at the regulator or water separator/filter should be checked first. A figure of between 6.2 bar and 6.8 bar (90 lbf/in² and 100 lbf/in²), which is the usual compressor cut-out level, should be available. Continuous supply can only be maintained if the main air lines are substantially larger than the take-off legs. In a large bodyshop a main line of 38 mm (1.5 inches) diameter would not be too large, with a short take-off leg to each regulator or water separator/filter unit. The compressor itself must of course be able to meet the total air demand from all users.

The size and quality of connectors and hoses used between the regulator, the breathing mask and the spraygun are all critical for good painting. The internal bore is the most important dimension and for both the hose and the connectors this should be at least 8 mm (5/16 inches) diameter. It is

Figure 116   Correct air pressure at the regulator is no guarantee of sufficient air during continuous spraying; use an air micrometer to set and monitor air pressure into the gun

possible that some HVLP sprayguns made need larger sizes. Good quality connectors and spraying hose will invariably comply with the minimum requirements. There are, however, many cheaper alternatives and, almost without exception, they are smaller. The other problem to be aware of is the internal collapse of cheap, old or carelessly used airlines. Tests at which the author was present established that there was a pressure drop of only 0.75 bar (11 lbf/in^2) on 12 metres (39 feet) of hose on an installation which conformed to the recommended basic guidelines.

Ideally, there should be only one straight-through connector and a length of hose not exceeding 9 metres (30 feet) connecting the breathing mask and the spraygun to the regulator unit. The use of 'Y' couplings should be avoided. The connectors on the mask filter may also cause obstruction to the free passage of air to the spraygun and so these, too, should be checked. If the spraygun is equipped with the air micrometer mentioned earlier, your checks are easily verified. Do remember, however, that you must hold the trigger on for at least 30 seconds to obtain a true reading of pressure drop and that the breathing mask should be working at the time of the check.

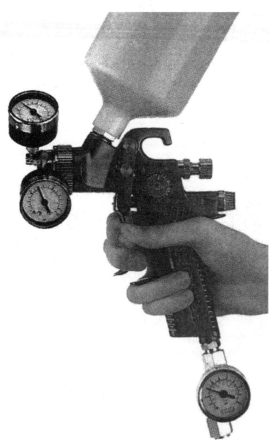

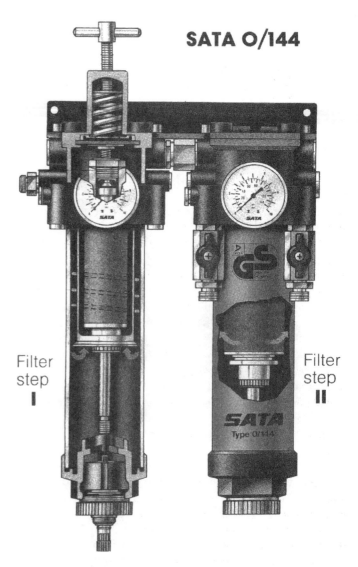

SATA O/144

Filter step I

Filter step II

Figure 118   Some sprayguns may allow a higher pressure than 12lbf/in^2 to escape from the air-cap at certain settings; caps are also available with two gauges to measure all air outlets and ensure compliance with the regulations

Figure 119   A sectioned view of a twin regulator

Types of materials

The products used for painting vehicles have undergone dramatic changes during recent years. The early mass production lines in the United States followed by the refinish industry relied upon quick drying nitro-cellulose. This material also became popular in the United Kingdom. In mainland Europe alkyd based paints were developed which, although more durable, needed baking.

Acrylics, the first true synthetic paints, have been available in various forms since 1955. Then the introduction of 'low bake', two-pack acrylics enabled bodyshops to reproduce the colour and durability of the 'high bake' production line finishes. Both factory applied and refinishing paints of this type give off solvents during drying and curing. These solvents are damaging to both human beings, and the environment at large, and consequently the paint industry has been forced to find alternatives. Legislation has brought about dramatic changes in many European factories and it is quite common now for new vehicles to be finished in water based paints. The refinish industry in the United Kingdom has also been changed by the EPA. This requires the gradual reduction of Volatile Organic Compounds (VOCs) to a very low level by 1998.

One type of paint which is permitted during the run-up phase to full implementation of the law is that known as a 'high solid'. This uses normal solvents but because, as the name suggests, there is a greater proportion of pigments and binders, there will be less solvent given off when it is curing. Solvent based paints are now available which will comply with the regulations beyond 1998 as currently published. Many paint makers are also offering a variety of water based systems for refinishing and these will be acceptable beyond 1998.

Two stages of the legislation have already been implemented, including the use of sprayguns that transfer at least 65% of the paint to the vehicle for all products except topcoats. This transfer requirement is usually met by HVLP sprayguns. The last stage is to come into force in 1998 when HVLP equipment must be used to spray all refinishing paints. HVLP sprayguns must be able to operate with a maximum pressure of 0.7 bar (10 lbf/in^2) at the tip. At the same time, solvent emissions must be reduced to the very low levels specified in the regulations.

Water based and high solids paints are the first results of the paint makers' research into alternatives and work continues on other possibilities, such as powder coatings. At the time of writing, it appears that water based paints are likely to be one of the means of complying with the regulations and meeting customers expectations after full implementation in 1998.

Specific recommendations are not possible with such a diversity of refinish systems. However, there are some general points which apply to most refinishing work. As far as possible, all the coatings used to produce a new finish should be from the same system to ensure compatibility. It is generally acceptable, however, to use stopper/fillers from other makers provided that they meet all criteria. This means that they should not interact with the paint coatings which will be painted over them and that they are approved for the substrate to which they will be applied. A vehicle manufac-turer's body warranty may impose conditions. For example, a particular anti-corrosion primer may be required or a special filler for the type of substrate. There are now many fillers suitable for galvanised panels but check that the particular example that you intend to use is both acceptable to the vehicle maker and will not react with the paint system. Aluminium, too, imposes special needs and a careful check should be made to ensure that the product is suitable.

Material and equipment storage

There has been a long-standing requirement for special storage facilities for refinishing materials. Traditional products such as nitro-cellulose are highly flammable and regulations calling for special storage conditions have been in force for many years. The flammability is perhaps less these days, but that has been replaced by the possible chemical interaction of one material with another and the highly toxic nature of some of them. It is still good practice then to keep paint materials in the special storage provided for them.

There will be a workroom adjacent to the spraybooth and this should only contain materials in use on that day. All other materials must be returned to long-term storage when they are no longer needed; this is another important example of good housekeeping.

There are other reasons, too. Materials for spraying should be kept at a temperature suitable for mixing and use, usually 20° Celsius. Some heavy build, spray-on fillers may be mixed by proportion but others need to be mixed to a particular viscosity. As the viscosity of any liquid varies with its temperature, the viscosity cup size and the time given will only apply at the stated measuring temperature. Keeping paints at this temperature avoids waiting for them to warm up.

Canisters of refinish materials should be kept closed at all times, except when actually pouring from them. This is to reduce the escape of fumes and prevent them from absorbing moisture. Some materials are described as being hygroscopic, which means that they absorb moisture from the atmosphere. Isocyanate activators are particularly bad in this respect and, if contaminated, may easily cause topcoat failures.

Mixing and spraying equipment must be kept in a place that is clean to avoid dirt contamination or repeat cleaning. This equipment should also be at the same temperature as the paints to prevent condensation forming on the surface of the metal.

Mixing and testing procedures

Detailed information on mixing all products, and especially viscosity figures for sprayed materials, must be taken from the maker's information. The maker's health and safety data sheets must also be consulted for directions on PPE and other measures. Here are some general guidelines that will help you to mix and test successfully.

All equipment must be clean. Solid body fillers should be mixed on a clean palette of non-absorbent material with a clean, slightly flexible spreader. This will limit the inclusion of air bubbles. For the same reason, use a 'wiping-in' action when mixing. Ensure that all the hardener is thoroughly spread throughout the mix. The colour should be even with no trace of streaking.

Spraying fillers must first be stirred to ensure that any settled solids are mixed. A mixing paddle in a pneumatic drill is ideal for this purpose. Mix no more than is needed and can be used during the pot-life time. Pour the filler into the mixing cup and top up with hardener to the correct level. Accurate measuring will ensure high quality. Then thin with the recommended thinner to the viscosity for spraying. Use only thinner from the same maker as the filler. Double check the viscosity cup size and timer setting. Be careful to synchronise activating the timer with lifting the full cup from the mix.

Application equipment selection and preparation

The spreader used for applying solid fillers should be of stiff, non-absorbent material but at the same time slightly flexible. It should be about as wide as a hand, permitting a steady wiping action to build and shape the filler in as few passes as possible. Perfect cleanliness is the only sure way of avoiding contamination problems.

Primers, primer fillers and surfacers must be applied by HVLP sprayguns. The air pressure at the cap on HVLP sprayguns should be checked periodically where this facility is available: correct aircap pressure is essential if paint is to be transferred at the correct rate. Spraygun selection and set-up should be done before mixing to give full use of the material's pot-life. Fluid tip, fluid needle and, where appropriate, air cap selection should be made to suit the material. These requirements are always given in the paint makers' information sheets or manuals.

Consumables such as chemical cleaners and tack rags must be to hand, together with all the necessary items of PPE.

Applying Fillers and Foundation Materials

Health and safety at work and PPE

Most of this topic was dealt with at the beginning of this chapter, on p. 75. If you have not yet done so, please read this important information.

You may be concerned about using infra-red curing equipment. This employs rays from the electromagnetic spectrum extending from visible light. There is no danger in looking at the radiant units when they are working. You should not, however, place any part of your body, especially bare skin, within the heated area.

Housekeeping and waste disposal

Much of this topic was covered at the beginning of this chapter and you should refer to it again if necessary.

Some points are important enough to repeat. One of these is the test spraying of sprayguns to confirm the set-up.

Under the regulations, it is not permissible to spray loosely into the air, even inside the spraybooth, when setting-up the gun for painting. It would seem reasonable then to carry out the first check onto masking paper set up for the purpose or into an enclosed cabinet. Final set-up, including test card spray-out (see Fig. 120), could also be done against a flat panel so that sufficient solids adhere to the surface and thereby comply with the regulations.

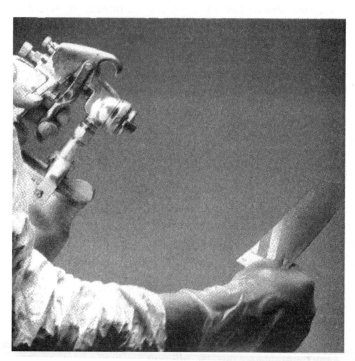

Figure 120   Painting the invaluable test card; keep it in your colour library

Many paint materials are not soluble in solvent once they are hard. It is important then to clean the spraygun immediately after use in the proper automatic cabinet.

Sources of information

The main source of information must always be the maker of the product that you are using. Other sources are given in Sources of Information at the beginning of this chapter, on p. 76.

Equipment and tooling

Much of this element has already been dealt with in the first section of this chapter, on p. 76. Your attention is drawn in particular to the topic of air supply and the requirements for painting, which is the most frequent cause of poor results during spray painting. It is even more important now that the transfer of paint is governed by regulations.

Drying or curing the primer filler is done conventionally by stoving in a low-bake oven at between 60° Celsius and 80° Celsius or by using infra-red drying equipment. Paint spraybooths, usually of the type called combi-ovens because they provide a spraying environment and an oven to dry or cure the paint, are being brought into line with the requirements of the EPA. By 1998 they will need to have very low levels of particulate emission (that is tiny particles of paint in the exhaust air) and have an equal or negative pressure relative to the surrounding atmosphere (to prevent the escape of fumes into the area outside). Lighting in the booth should be by 'daylight' lamps and be even and without shadow.

Infra-red equipment is available in a range of sizes from small hand-held units, suitable for panel warming and spot repairs, through to twin booth, computer controlled and automated versions for high throughput paintshops. They all work by using the infra-red waveband of the spectrum which is immediately above the zone of visible light. The principal benefit of infra-red is the greatly reduced curing time. This can range typically from 2 minutes for polyester filler to 10 minutes for topcoat. A full respray may be cured in as little as 15 minutes. One peculiarity is that colour determines curing time; light colours reflect the rays so they need a little longer curing time.

Application

Here we are concerned with the sprayed application of primers and primer filler coats. Some of these fillers may be of high build type to improve shape and contour. Other coatings are primers and normal spraying primer fillers.

Bodyshops vary in the way that filling and sealing activities are organised. It is not unusual for priming to be carried out in the paintshop before the vehicle is returned to the panelshop for completion of initial stopping, shaping and sealing. I have adopted this procedure and you will find information on the use of stoppers or spreading fillers and sealing in Chapter 3, together with some aspects of shaping the repair.

Much of the foundation knowledge needed for spraying foundation coats also applies to topcoats, except for basecoat and clear finishes. For that, and for masking, you must consult Chapters 6 and 7.

Paints are going through a period of change and development as a result of the EPA. This sets out a timetable for limiting the discharge of VOCs into the environment by insisting on a number of measures. These include, amongst others, the proportion of sprayed paint that adheres to a surface and the amounts of VOCs in paint products. Here are the requirements as they affect you:

- Primers and primer fillers must be applied in a manner which ensures that a minimum of 65% of the paint adheres to the sprayed surface. To meet this requirement you must use an HVLP spraygun correctly.
- Topcoats will be of the type called 'high solids', that is to say that they have a higher proportion of pigments and binders to solvent than ordinary paints.
- From 1998 topcoats must have a very low VOC content. Some are likely to be water based.
- All finishes must be applied with HVLP sprayguns from 1 October 1998.
- HVLP sprayguns must be operated with a tip pressure of not more than 0.7 bar (10 lbf/in^2).

There are differences in using HVLP sprayguns compared to normal types and these will be obvious during practical training. One factor is clear – they demand careful control and preparation. The basic techniques that will obtain high quality 'from the gun' finishes are much the same, whichever paint or type of spray equipment is used.

The gun itself must be in first class condition as far as the working parts are concerned. Guns should be carefully cleaned at the end of every session to avoid the problem of cured paint in jets and passages. There will be times when blocked or dirty jets must be cleared and on these occasions profiled cleaning needles from the spraygun maker must be used. The cleaning brushes supplied by the makers may also be needed. Always take care that the matched air cap, fluid nozzle and needle are kept together.

Routine maintenance of your sprayguns is another important part of high-quality painting. Faulty spray patterns, intermittent spray and spitting are just some of the problems that can occur if the gun is faulty. Renewing seals, lubricating components and adjusting settings are all routine operations that must be done regularly if you are to have trouble-free operation. Make sure that a set of spares is available for each gun.

The materials that are supplied to the gun are also critical to good painting. Enough clean air at the correct pressure is essential. This topic was dealt with under the heading of Equipment and Tooling earlier on p. 90 of this section. To confirm the available air pressure an air micrometer, and possibly also an air cap gauge, are essential. The air micrometer is for both types of sprayguns and the air cap gauge for HVLP.

The other material, the paint to be transferred, is equally important. Thorough mixing of the pigments and binders is vital if the material is to spray, flow and cure to a good finish. Good opacity, the ability to hide the coat below, depends on well mixed paint. The separate components of two-pack products must be mixed in the correct ratio if they are to cure properly and wear well. Topcoats are kept on the mixing machine and its care and use are dealt with in Chapter 7.

The degree of thinning determines how well all materials spray and the quality of the finished job. The use of the correct viscosity cup and an accurate timer in the right temperature enable you to check that the mix is a suitable consistency for spraying (see Fig. 121).

Cleanliness is of paramount importance, both for the appearance of the finish and also for the smooth transfer of the paint. The paint container must be spotless. Always filter paints into the paint container to remove any solid particles that could interrupt the smooth flow of paint or even disfigure the surface (see Fig. 122).

When applying foundation coats that will be sanded, take a pot of very thin black (or other contrasting colour) into the booth to use as a guidecoat for flatting. It will also serve to partially clean the gun passages. Guidecoats may be applied wet-on-wet.

Failure to follow any one of these guidelines could spell disaster so here is a summary of these vitally important points:

- Observe all health and safety and EPA requirements.
- Keep materials at 20° Celsius or above and in closed canisters.
- Mix paint ingredients thoroughly.

- Measure ratios of two-pack with a mixing stick and mix well.
- Always use ingredients from one maker, including the thinners.
- Use a mixing stick when adding thinner and check with a viscosity cup and timer.
- Always use a disposable filter when filling the gun canister.
- Throughout the process clean everything, all the time.
- Never mix makes or systems.

Figure 121 Correct paint temperature and mixing are two of the foundations of good refinishing; another is to check the viscosity and correct it if necessary

Figure 122 Always strain paint unless the maker specifies no filtering or straining

Drying and curing

The difference between these two processes is simple. Drying depends upon a solvent evaporating away. Curing involves a chemical action between two components of the two-pack products. This gives rise to a 'pot-life', the time during which the product must be used. Solvent will also need to evaporate away from two-pack materials if it has been added prior to spraying.

Nitro-cellulose and acrylic paints air-dry 'dust-free' relatively quickly but two-pack paints which can be air-dried have the disadvantage of attracting dirt and dust because of their longer drying time. They can be made so that they air-dry quickly, but then there is little time to apply the paint neatly and for it to flow into a smooth film. Heating a surface newly painted with two-pack paint is the only way of achieving both workability and quick drying.

The heated spraybooth has been the main way of speeding up the drying or curing of paint for many years. It has been developed to an advanced state where lighting, airflow and filtration provide an almost perfect working environment. The best booths can be at working temperature within minutes and, by recirculating air, keep costs within reasonable limits. Like any other equipment they will only be as good as their users keep them, of course. Regular cleaning, filter changing when necessary and normal routine maintenance are essential if a booth is to give the best results.

An important consideration under the EPA is that the air pressure inside the booth must be slightly less than outside at −5 pascals. This is called negative pressure. This does mean that if there are any gaps around the doors, and the outer area is dirty, there is a risk of dirt entering the booth. This law has followed a number of cases where people working outside a booth with positive pressure have been sensitised to isocyanates. This is not surprising as it was once thought desirable that the pressure should be higher to prevent the entry of dirt.

The drying or curing equipment that has improved productivity by reducing the time spent on low bake is infra-red (see Fig. 123). As with conventional booths, there will be specific instructions for each make of unit. There are a number of general points to observe if the best results are to be obtained:

- Care must be taken where plastic masking is used. If there is any doubt as to the suitability of a particular make or type of masking for infra-red curing, it is best to be cautious and pull it back from the treated area.
- Substrate temperatures do not generally rise beyond 80° Celsius although there are some colours where they may reach 100° Celsius. If there are electronic components either against or very close to the panel it is as well to move them away. If you are concerned about the possibility of excess panel temperature, this may be checked with a suitable industrial thermometer.
- The generally accepted distance for most infra-red emitters is approximately 0.8 metres away from the painted surface.
- Where infra-red is used only those parts of the body exposed to the rays will be cured quickly.

■ Once drying or curing is complete, take care to observe any additional cooling time before reworking.
■ If problems do occur, discuss them with both the paint and the equipment makers as either may have the answer.

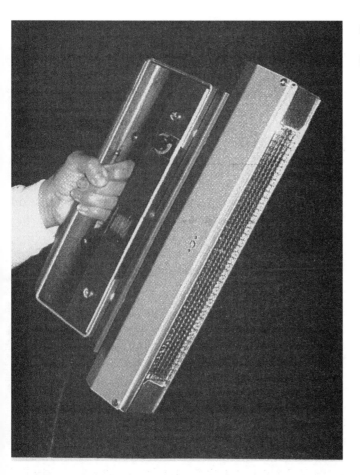

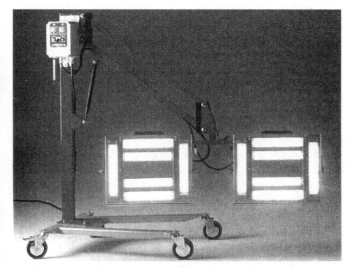

Figure 123 (a) A hand held infra-red heater and (b) multiple units on a stand

■ If the panel of emitters does not fully cover the newly painted area, the curing operation will have to be repeated. Should that be necessary, some of the time saving gained from using this equipment will be lost and conventional stoving may be more appropriate.
■ The product is mixed and applied with infra-red curing in mind. Only use products and infra-red when both the product and the equipment are known to be compatible with each other.

Material and Equipment Preparation

Health and safety at work and PPE

The hazards facing the painter are twofold; the fumes given off by paints and solvents and the minute particles of paint in the spray mist. The highly toxic nature of some materials, particularly hardeners, is giving way to more 'user friendly' ingredients as a result of the Environmental Protection Act (EPA). The methods of applying paint are changing, too, for the same reason. Despite these changes there will continue to be risks associated with a painter's work and the correct use of the appropriate PPE will continue to be essential to maintaining good health. For information on the general hazards arising from the spraying of paint, refer to Chapter 6.

Paints containing isocyanates are likely to be in use for some time. The danger associated with using these products stems from the tiny amounts that can affect your health. Just 2 hundredths of a milligram in a cubic metre of air, that is much less than one part in a million, is all it takes to have an effect. As a result of exposure the body becomes sensitised to isocyanate. Thereafter, the slightest trace of the solvent in the air that is breathed will result in a form of asthmatic attack. There are over 1000 new cases every year of people made asthmatic through their work.

The first signs of sensitisation are runny eyes and nose and also itchy eyes and nose. These signs usually appear during the evening and also in the night. Because of this they may not be thought to be work related. If there is noticeable improvement at week ends or particularly during a holiday, this indicates a work related sensitisation. More severe symptoms of sensitisation are wheezing, tightness of the chest, breathlessness and coughing.

The most important point to remember is that you must take every measure to prevent these symptoms occurring. If they do, they are with you for life!

Here are some additional points about personal hygiene:

Always wear solvent proof gloves when handling paint. Accidental splashes of paint on the skin should be wiped away promptly and any residue removed with a suitable hand cleanser and washing. Skin which has been in contact with peroxide, acid and organic catalysts should be washed with a 2% solution of Sodium Bicarbonate. Apply some skin-care cream afterwards to restore the protective substances.

Other ways in which you may care for your health will, no doubt, occur to you. It is a valuable possession which can rarely be recovered once it is lost. Many of these points apply to others around you. The most important of these must be the accidental sensitisation to isocyanates of others outside the booth because of leaks. If you know that your booth has positive pressure, report it to your supervisor immediately. Likewise, if there is a broken door seal or other defect, this too should be reported at once. Needless to say, spraying should never be done outside the booth.

Housekeeping and waste disposal

The disposal of waste is dealt with in all three sections of Chapter 6, which you should read if you have not already done so. In this chapter the use of topcoats is the main concern and here the principal waste will be the disposal of unused paint, gun cleaning residues and used, disposable masking.

It is worth mentioning again the self-ignition risks of certain types of paints when they come into contact with sanding dust or other waste. The metal bins used for dry paint waste should always be damped with water and covered. Liquid waste must be put into the metal containers provided for that purpose. If your bodyshop has a reclaiming unit make sure that you put into it only those materials that it is designed to handle.

Apart from complying with regulations, good housekeeping is fundamental to successful painting. For many

Hazard	Protection
Damage to skin tissue from contamination by paint or solvent	Use a barrier cream to maintain skin moisture Wear solvent proof gloves when handling paint and equipment Wash your hands before using the toilet as well as after
Ingestion, that is taking in toxic substances through the mouth	Wash your hands before eating or drinking, even when using a cup Do not drink or eat in the workplace: it is one of the easiest ways of damaging your health and is illegal

painters, this is a matter of self-discipline as they work on their own; this is also an advantage as there is no one else to make a mess. The condition of the paint cans, mixing machine and scales can make or mar a finish before you have even lifted a spraygun. Booth cleanliness, as well as that of the surroundings, is yet another part of high-quality painting. Not least is the condition of you, the painter. Are you wearing non-fibrous overalls and are they clean (in terms of loose fluff and dust)? All these items are those of routine care, carried out day by day. If that care is exercised, then high-quality paintwork can also become routine, day by day.

Sources of information

The source for most of the information that a painter needs must always be the maker of the paint system that is being used for the job. There are exceptions and it may sometimes be necessary to consider what should be used on a vehicle that has a body warranty, for example. All paint makers provide a comprehensive information service, which may also be linked to some items of special equipment, such as a colour computing system.

All the information on the paint systems used by the bodyshop is very valuable and should be kept clean and secure. Where up-dating information is supplied this should be acted upon immediately to prevent problems. The mixing system data, which is usually on microfiche, must be maintained in clean condition and in the correct order. When using a fiche, write the mix on the back of a spray-out card and replace the fiche immediately. The card can then be taken to the bench for mixing and subsequently used for the test spray-out.

These spray-out cards form the basis of one of the most useful aids that a painter can develop for personal use. A card should be prepared for the topcoats on every job and completed with details of the vehicle. If these cards are then kept in a suitable file or box, they will form an invaluable reference of colours and shades used on the vehicles that have been painted in that bodyshop. This will be of great value for future work on the same vehicle and on others of the same colour. Vehicle makers' shades vary a great deal and such a file built up over a period will become extremely valuable to a painter. If you know that you have a similar shade on an existing card, check it and, if it is right, use it.

Other aids are available from the paint makers such as colour swatches and lists of all vehicle makers' colours. These usually give such information as makers' colour codes, years of production when used and the availability of a formulation.

Refinishing materials and paint systems

Makers of mass produced vehicles use finishes which give a durable surface with good colour retention. The paints are designed for production line painting and cure to a hard finish. This has the double benefit of less chance of damage when the vehicle is being built up and good wear resistance for the owner. They are cured at temperatures of up to 150° Celsius and are known as 'high stoved' or 'high bake' finishes.

The materials that you as a refinisher use are designed for either air curing or low stoving or baking at no more than 80° Celsius and are known as 'low bake' finishes. These lower temperatures are necessary to avoid damage to the assembled vehicle. Even the protective waxes which are flooded or sprayed into the internal cavities of many cars will not melt at the lower temperature.

Paints are mixed from three basic ingredients; pigments, binders and solvent:

■ Pigments provide the colour in topcoats, the building of a good layer in fillers and anti-corrosion qualities in primers.

Job	Source of information
PPE selection	PPE makers' guidance notes and sales brochures Specialist suppliers' information Product makers' data sheets Vehicle makers' information MIRRC (Thatcham) information
Selection of materials	Company standing instructions Job card or estimate Vehicle makers' information MIRRC information
Equipment selection and set-up.	Paint makers' information Equipment makers' information Suppliers' information Training material
Paint mixing	Paint makers' microfiche or manuals Colour manuals Colour library

- Binders enable the paint to form a film and adhere to the substrate by binding the particles together.
- Solvents are used to make the paint fluid and to thin it further for spray painting.

Refinish materials can be grouped together in the way that these constituents change to become a hard coating. The groups are lacquers, enamels and two-pack materials.

- Lacquers dry by the evaporation of solvent. Such a paint is cellulose, once very popular for refinishing in the United Kingdom. American paint classification may also include some of the next group in this category.
- Enamels are a complex group with both air drying and low bake finishes. Both types dry in two stages. The first stage is solvent evaporation, which is then followed by a chemical reaction. The low bake finishes need heat to start and complete the chemical reaction.
- Two-pack materials are in general use throughout the refinish industry, from polyester stoppers and fillers through to the topcoats and clear lacquers for final finish. This group is also now quite complex with the addition of new products that comply with the Environmental Protection Act (EPA).

Painters are experiencing many changes in the products that they use. The technology of paint is developing at such a fast rate, mainly in response to the EPA, that more changes are likely in the foreseeable future. Whenever a new product is introduced, the maker issues data sheets to explain how to use it and runs training courses from which you can gain hands-on experience. The only way in which you can become, and remain, an expert is to use the information and attend for training.

Two-pack materials now cover such a wide range that it may be helpful to put the products into some sort of order with some of the factors most have in common:

- Stoppers and fillers of the polyester type are mixed with peroxide catalysts as hardeners. Spreading, brushing and spraying body fillers have powders such as talcum instead of pigments and resin to bind them all together. The peroxide based hardener causes a chemical reaction that hardens the filler. The hardener must be in the correct proportions to the filler to ensure that 'unused' chemicals do not cause bleaching of the subsequent coats (see Plate 13). The working time or 'pot-life' of some of these fillers may be only a matter of minutes.
- Primers provide a base for the succeeding coats. Their primary purpose is to ensure good adhesion of the paint film to the substrate. They may also seal any underlying coat from those applied afterwards. Self-etching primers contain acid which eats away (etches) any film on the substrate surface to ensure a good key. Some vehicle makers with long-term body warranties may insist upon the use of special primers to maintain the warranty cover.
- Filler coats, sometimes combined with a primer, provide a high build foundation for the colour coats that follow. It is sanded to smooth out minor imperfections or irregularities to provide a perfect surface.
- Colour coats themselves fall into a number of groups. Solid colours are those which are not translucent. Metallics are translucent with aluminium particles to reflect the light. There are also the pearlescent and mica variations of metallics. All these finishes except solid colours must be lacquered as they are not naturally glossy. Solid colours may also be lacquered to provide additional lustre and depth. A matt topcoat followed by a lacquer is called a 'basecoat and clear' finish.
- Thinners are the other all important ingredient in the paint system. They are available in a number of forms and should be used strictly in accordance with the paint maker's recommendations. Those that should be used will blend perfectly with the paint itself. The evaporation rate will influence how the paint flows and the drying time.

There are now a number of terms applied to paints which have arisen from the research into products that comply with the EPA, that is paints that do not contain more than a specified amount of Volatile Organic Compound (VOC). These paints are in two main groups; water based and high solids:

- Water based paints are those that use water instead of solvent to make them workable. New vehicle production has been gradually changing to water based products for some time to overcome the high solvent emissions from the factories. Water based materials for refinishing have followed more recently.
- High solids materials use VOC solvents but the content is greatly reduced. Such a paint may contain up to 70% paint solids compared to conventional two-pack with about 35% of solids.

Apart from limiting the VOC part of paint the regulations also call for the reduction of 'particulate emission', that is the paint solids escaping to the atmosphere from the booth chimney, to be as low as 10 milligrams in each cubic metre of air.

Equipment and tooling

The microfiche reader must be kept clean to avoid misreading formulations. Bulbs can also become blackened in use and it is sometimes necessary to change them before they burn out. An alternative to the fiche system is the use of a computer linked to electronic scales. The paint reference code is used to initiate the mix and give the readouts required on the scales. You will appreciate that sensitive electronic equipment must be housed in conditions that will not give rise to damage and it must be used in a careful manner.

Scales are of critical importance due to the precise readings which are taken. Keep them clean and calibrated. Of equal importance is the condition of the canisters of mixing colours on the mixing machine. Keep the pouring spouts clear of paint build-up to allow accurate, one drop measuring.

The mixing machine itself is the basis of good colour matching. This can only be achieved if the base and tinters are kept thoroughly mixed. If measuring jugs are used for mixing they must be spotlessly clean. New cans are often preferred but these too must be kept in a clean condition. The same is true for viscosity cups, particularly through the measuring orifice. Mixing sticks will not be easy to read if they are not clean.

Information on sprayguns and their care will be found in Chapter 6, together with important issues concerning air supply. The quality and quantity of air supplied to the booth and then to the spraygun is fundamental to the successful application of topcoats.

The use of a spraybooth and separate oven has largely given way to the combination booth/oven for body refinish, where vehicles are usually sprayed and cured in the same cabin (see Fig. 124). Hot air curing may be supplemented by the use of infra-red units which dramatically reduce the time needed. In the highest throughput installations, an auto-mated infra-red arch will be used in alternate booths: while one vehicle cures, another is painted. Then the situation is reversed: the arch is moved to the other booth for curing while spraying is done in the first booth.

Other details about spraybooths and infra-red units can be found in Chapter 6.

Material and equipment storage

The temperature at which paint products are kept is impor-tant for ease of use and application. Most importantly it is fixed between 20° Celsius and 25° Celsius so that accurate viscosity readings may be taken when mixing paints with thinner for spraying. When cold they will be too viscous and as a result are likely to be overthinned. New cans of paint may be kept in lower temperatures but it is important that they are allowed to rise to room temperature over a period of 24 hours before use (see Fig. 125).

Base colours and tinters are kept ready for use on the mixing machine. The machine must be switched on and run for at least 10 minutes, twice a day. It must also be in a room kept at not less than 20° Celsius. When a new can of base or tinter is added to the machine it should first be thoroughly stirred by hand or with a rotary mixer (see Fig. 126), until all the pigment is distributed and none remains as sediment in the bottom. Keep the pouring spouts and stoppers of canis-ters clean. Tinters must also be stirred thoroughly before **every** use.

It is also important to keep equipment such as sprayguns, viscosity cups and mixing sticks at room temperature as there is a risk of condensation forming if they are cold. The same is true of cans used for mixing, of course.

Mixing and testing procedures

A respirator that can handle small quantities of toxic materi-als is adequate for mixing room use provided that replace-able elements are changed in good time. Materials containing isocyanate may only be sprayed when the pain-ter is wearing an air-fed face mask and in a closed spray-booth. Isocyanate may also be absorbed through exposed skin so you should always wear full protective clothing.

As with foundation coats, the mixing of topcoats is basi-cally a two stage process; the paint is mixed to the correct formulation and then thinned to make it suitable for spray-ing. Much of this topic is covered in the element on the

Figure 124 (a) Buses and (b) railway coaches also need spraybooths

Figure 125 Water based finishes must be put into lined cans

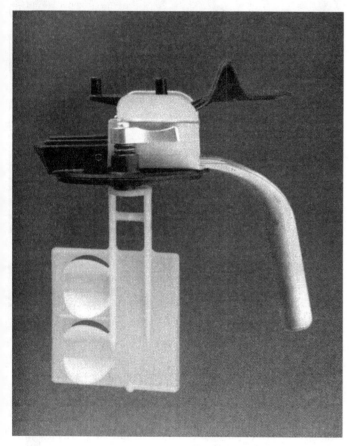

Figure 126 Mixing machine agitators must be of suitable type for water based finishes. Delay will cause the colour to look quite different.

application of foundation coats in Chapter 6. The great difference with topcoat is that colour matching is now a major concern. To verify colour and reach a match there are a number of things that must be done:

- Clean the paintwork of the vehicle thoroughly to remove traffic film and restore colour.

- Check the comparison in natural North light, under artificial daylight lighting or a specified light source.
- The surroundings must be of a matt, neutral nature to avoid coloration (apparent tinting of the colour by the surroundings).
- The checking sample or test card must lay flat on the surface for true comparison.
- The angle of view should be straight on for both solid colours and metallic face tone. Metallic side tone is checked at 45°.
- The limitations of human eyesight change colour perception very quickly. Colour comparison must be brief.

Most painters prepare the colour stage in advance so that the first spray-out card is painted at filler coat stage. This means that final tinting can be done while the filler coat is drying. It also has the benefit that if the opacity of a topcoat colour is not particularly good, there is still time to apply a tinted filler coat to give body to the topcoat colour. Keep a record of the mix each time that you make an adjustment. In that way you will be able to mix precisely the same shade again. Once the colour is fixed, identify the card and file it in your personal colour library.

Colour

You may have noticed that it is quite rare for any two people to agree entirely about a range of colours. We rarely agree about blues and greens in that area where they blend from one to the other. That is because our perception of colour is decided in our brain; we do not actually see it. What the eye senses and conveys to our brain is the wavelengths of the reflected light that it can detect. Moreover, it can only sense three – red, green and blue. The other 'colours' are made of a mix of these three.

White light consists of seven colours: red, orange, yellow, green, blue, indigo and violet. These are the colours of the spectrum. You can memorise them by the phrase 'Richard of York goes back into Venice'. Each colour occupies a small range of wavelengths, the shortest is violet and the longest, red. As light shines on a painted surface some of the light 'waves' are absorbed by the pigments in the paint. The light waves left over are reflected. Our eyes sense the amount of red, blue and green in this reflected light. This information is fed to our brain, which decides what the colour is. If you look at the same object under totally different light, like the orange sodium-vapour street lamps for example, they reflect quite different wavelengths and we 'see' a different colour.

The three primary colours of light make white light without using the others in the spectrum. Watch the red, blue and green spotlights on a stage. This is called additive mixing because when mixed they become white.

The colours of pigments react in a totally different way. Mixing pigments together will also create different colours but the primary colours are magenta (a red), cyan (a blue) and yellow. Mixing these primary colours together produces black. This is called subtractive mixing because all the light waves are absorbed, none are reflected.

The quality of a colour is described in a number of ways. Unfortunately, not everyone agrees on what each aspect is called. Here they are with some alternatives:

■ *Hue, tint, or tone* – the colour.
■ *Saturation or chroma* – purity of the colour or degree of blending with others.
■ *Brightness, clarity or value* – how light or dark it is.

These are the additional descriptive terms used for the reflective effects of metallic finishes:

■ *Face tone* – the appearance when viewed at an angle of 90° to the surface.
■ *Flip or side tone* – viewing the surface from 45° or less.
■ *Mass tone* – the dominant colour.
■ *Under tone* – a less prominent colour.

Metallic finishes rely for effect upon the reflection of light from both the coloured pigments and the particles of aluminium contained within the translucent paint. Every aspect of mixing, thinning and spraying a metallic finish will influence the final appearance. The mixing of the colour is similar to that for solid colours except that some tinters do not affect face and side tone to the same degree. Aluminium flakes are added to provide the additional effect. The flakes may be available in different sizes and they will have varying effects upon the reflection and so seem to alter the colour. The viscosity of the paint is critical. These are the results of variations in thinning and also in applying correctly thinned paint in different ways:

■ Too thin and the flakes will settle to the bottom of the film.
■ Too viscous and the flakes will be held firmly as they land, with little settling.
■ Correctly thinned but applied in thin coats produces flakes parallel to the panel surface. There is pronounced colour variation between face tone and side tone.
■ Correctly thinned but applied heavily will result in slower curing and more time, allowing the flakes to settle in a random fashion. Multiple flake angles result in strong, multi-angle reflections. There will be little difference between the face and side tones.

The painter can totally control the appearance of the finish by mixing in different proportions and thinning. Gun settings, speed and distance also influence the finish. These are the reasons why two painters can obtain quite different results from the same paint mixture. It is also the reason why the utmost care must be taken in assessing the original finish in the first place, and adjusting colour, thinning and application to match the existing paint film. It is also vitally important that no judgement is made about the spray-out test cards unless they are fully finished with a clear lacquer.

Topcoat Application and Curing

Health and safety at work and PPE

The major risks have already been mentioned, but bear repeating. The particles of paint that form the spray mist are of a size that can block the airways in the lungs. When non-toxic paint is used the respirator must be able to filter out these unwanted particles. Many of the paints currently in use contain isocyanate, which can, in very small quantities, sensitise a breather . Only an air-fed face mask provides adequate protection from paints containing isocyanate. It is worth considering using 'peel-off' protection to prevent permanent damage to the visor from paint. Chapter 4 and Chapter 6 contain other vital information on protecting lungs and, if you have not already done so, please read them now.

Hands should be protected by barrier cream and by wearing gloves resistant to solvents. Overalls should be disposable, paper based type (see Fig. 127). These will not become affected by static electricity nor give off loose fibre to cause inclusions in the wet paint.

Figure 127   A painter correctly protected in the mixing room

Housekeeping and waste disposal

The first section in this chapter dealt with the main issues of housekeeping and waste disposal. The only additional point that is relevant at this stage is the test spraying of sprayguns. Sometimes, when visiting bodyshops, I have been disappointed to see multi-coloured walls in the spraybooth or mixing room where the spraygun has been tested and adjusted. An accumulation of old masking, a dirty mixing table and other rubbish can also be seen too frequently. This is not only dangerous and illegal but depressing for those around. The painter is, above all, the one craftsperson in a bodyshop who must be fastidious over cleanliness and good order. The success of your work depends upon it and will inevitably show through in a job well done.

Information sources

Refer to the first section in this chapter for information sources on paints and painting.

Equipment and tooling

Refer to the section at the beginning of this chapter, pp. 97–8, on equipment and tooling.

Painting conditions

Sufficient, clean air is one of the most important needs for high-quality, trouble free painting. Even though the latest sprayguns work well on lower pressures, there must be more than enough volume to supply the gun and the air-fed face mask. You will find the information that you need to check out your own installation in Chapter 6.

The conditions for mixing and loading paint included such topics as cleanliness and temperature. These, of course, are just as important when it comes to applying the paint. The general cleanliness of the booth is very important. Masking, masking tape and other materials laying about can harbour dust and potential inclusions. Whether they are old or new, such litter should be removed at the earliest opportunity. It follows that materials such as masking should be kept as clean as possible during storage to minimise the dirt and dust that will be carried into the booth with them. The walls and lighting units should be as clean as possible to provide the highest and most even level of light. The panels and glasses should be cleaned regularly to remove any dust or spray accumulation.

The temperature of both the booth airflow and of the component or vehicle to be sprayed is important. A film of condensation can easily form if the temperature of the work is not at the same level as the booth air. Allow sufficient time for all temperatures to stabilise at around 20° Celsius to 24° Celsius before commencing spraying.

The circulating air flow is designed to follow gravity, coming in through the ceiling filters, flowing over the vehicle, and leaving through the floor grills. Fans force the air through at speeds as high as 36 metres per minute (120 feet per minute). The booth air is usually changed completely between 4 and 6 times every minute. Where an infra-red arch is installed, a movement as high as 8 times per minute may be beneficial. Such a high volume of air flowing through the ceiling filter calls for good standards of maintenance to ensure 'from the gun' finishes that need no further attention. The filters must be of good quality and be changed whenever necessary. The hinged panel in which they are held must be a tight fit. If there are gaps, air can by-pass the filter and carry dirt inside (see Fig. 128).

The use of water borne paints to comply with the requirements of the EPA has brought about a need for some changes in drying. During the initial drying phase before baking, the surface air becomes laden with moisture and cannot accept any more from the paint. This occurs despite the high speed of the airflow over the paint. This slowing

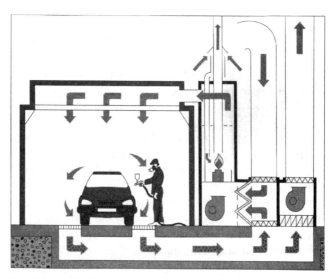

Figure 128 Circulation and extraction through a typical spraybooth

down of the 'flash-off' phase can be overcome in three ways. The air flow may be increased by modifying the booth or by using air blowing guns. Secondly, the air temperature can be raised. The third method is to use infra-red, which dries the paint by heating the substrate. As previously mentioned, a high rate of air flow is desirable with infra-red.

The EPA lays down requirements for spraybooths that will come into effect, in stages, between now and the year 2000. To comply with the EPA, spraybooths must be fitted with paint particulate filters to prevent small particles of paint being exhausted to atmosphere. These create an obstruction to the flow of air leaving the booth. Another requirement of the EPA is that no paint or solvent must escape from the booth into the surrounding workshop. This means that the air pressure in the booth must be kept a little below the atmospheric pressure outside. To cope with these requirements, fans are installed to draw the air through the outlet filters. The input and output fans are linked to control the slightly negative pressure. These output filters must also be examined periodically and changed whenever necessary.

The reason for this is that there is also an alarm to warn of booth overpressure which can close the system down almost at once. If this happens during spraying there is likely to be considerable extra work involved in preparing the vehicle for eventual completion. The makers of spraybooths invariably prepare a maintenance scheme and following this should avoid problems. It is also in the best interests of good quality work. A reduced airflow will almost certainly result in turbulence in the booth during painting. In extreme cases this will cause semi-dry particles of paint from the free spray to be deposited onto the surface of completed paintwork.

To help you avoid pressure problems booths are, or will be, equipped with a pressure gauge on the control panel, which must be checked by the painter before any work is carried out in the booth. It will be marked in Pascals, which are 100 000ths of a bar. The reading should be less than zero and not usually lower than −10 Pascals. You must make it a routine to check the booth pressure whenever you enter it to do painting.

Masking

Masking is an art in itself and is just as important as the mixing of paint and spraying the bodywork. It is a part of the job that the average customer never sees, and yet the evidence of good or bad masking can be seen by anyone. The lack, or the presence, of overspray, sharp edges and smooth blending tell the world how good you are. The purpose of masking is to protect those areas that do not need paint, and to provide those parts that do with suitable edging to help the painter to finish an invisible repair.

There is no excuse for not creating excellent masking as the painter today is spoilt for choice by the wide range of materials available. The traditional masking paper and tape is often used for much of the masking. However, there are also different, and often quicker and easier ways, of masking parts of the repair. These can help to give a high standard of finish. This topic was covered at some length in Chapter 6. Here is the checklist of important points to remember when masking:

- Paper should be strong, non-absorbent and designed for the purpose.
- Use a dispenser to attach tape as you tear off usable sized pieces.
- Check the type of plastic sheeting if that is what you use. Some are affected by infra-red rays and, if you use this form of heating, it may need to be pulled out of the way.
- Liquid masking may be the answer for your work. Check that it will withstand the processes that you intend to use.
- Masking tape should be strong and have a pronounced crepe effect.
- You will need both narrow and wide tape to mask off trim and define edges. Wide tape is also needed to close off gaps.
- Bodywork and trim must be at or above workshop temperature if the tape is to adhere properly. If the surface is even slightly cooler a film of condensation can form that will prevent adhesion. A hand held infra-red heater is ideal for quick, gentle warming.
- A sharp edged wooden or plastic spatula is ideal for 'tucking-in' edges into tight corners.
- Use the special tapes and inserts to lift the flanges of flexible trim. This allows any fade-out to be hidden.
- Roll masking or shaped foam strip should be used for all edges that need to be blended. Door pillars can be provided with a larger hood to extend the fade-out area.
- Roll masking may also be used as an addition to edge masked trim. The roll will give a blending effect towards the trim. The roll can be removed for the final coat to allow the paint to flow without creating a significant edge.
- Cover all openings or gaps where dust could be present. This will prevent it blowing out or overspray from penetrating into the opening.
- Use wheel covers and mask wheel arch openings.
- Provide adequate protection for the underbody.
- Raise the bonnet and mask the engine compartment independently of the outside masking when painting at the front.
- Make stiff flaps for panels that must open.
- Keep it neat to avoid dust traps.

Good masking takes time and needs patience but it will be repaid by the absence of unwanted paint and especially by the fine finish that can be the result.

Spraying methods

New vehicles are mainly painted by an electrostatic process where the body is given an electrical charge and the paint is charged at the opposite polarity. Paint is fed under pressure to sprayheads revolving at high speed, typically 20 000 revolutions per minute or more. As the minute droplets of the atomised paint emerge from the sprayheads they are attracted to the body in a reasonably even coating, although the roof, bonnet and the tops of the front wings usually have greater depth than the sides. The main reason for this is that the sprayheads are positioned above the body. The paintwork is then cured at a temperature of up to 150° Celsius, producing a finish known as 'high bake'. In contrast, refinishing is typically stoved at 80° Celsius or below and is called a 'low bake' finish.

Such equipment is far too complex and costly for the body repairer and the stoving temperatures are only suitable for stripped out bodies. It should be noted here that factory rectification work carried out after assembly is done in the usual bodyshop way using low bake materials. This will also apply to transit damage repairs. You will find full information on recognising substrates in Chapter 6.

All other methods of paint spraying involve a painter using a spraygun. Medium volume production may be finished using a pressurised paint supply from bulk storage. Low volume production finishing, or those refinishing large areas such as commercial vehicles, will probably use a pressurised paint supply from a canister. Conventional small vehicle refinish only needs a 1 litre pot or even less. These will be of the traditional high pressure spray type or the later, high volume, low pressure (HVLP) design.

Conventional sprayguns use what is known as the Venturi Principle to mix paint with air and spray it in tiny droplets. Air rushing through a reduced space speeds up, just like the rapids on a river. In our spraygun this speeding-up creates a depression, an area of low pressure, at the face of the air cap. This is where the paint outlet is placed. The atmospheric pressure all around us is also pushing down onto the paint in the spraygun cup and paint is forced into the gun. When the needle is pulled back by the trigger, paint can flow out of the nozzle into the depression. The paint is atomized in the air stream. When the trigger is released the needle closes the nozzle, shutting off the supply of paint.

There is just one snag with the traditional spraygun. It blows compressed air at the workpiece at up to 5.5 bar (80 lbf/in² – pounds force per square inch) and at a rate of between 283 litres and 425 litres (10 cubic feet and 15 cubic feet) every minute. The paint is carried in this air stream and, because it is travelling at high speed, about two thirds of it bounces off the bodywork and is lost. Only some 33% or so of paint is transferred to the body. This is called Transfer Efficiency (TE).

Legislation in the form of the EPA has brought about a rapid adoption of the spraygun that overcomes much of this disadvantage, the HVLP spraygun. The TE of this type of spraygun is in excess of 65% as judged by independent tests. How is this brought about? The secret lies with a device incorporated into the spraygun that atomises the paint in a similar way but then carries it out of the nozzle on an air pressure of no more than 0.7 bar (10 psi). However, the gun still needs at least 425 litres (15 cubic feet) of air every minute to operate.

Paint cups mounted above the gun are called gravity-fed and are usually no more than two thirds of a litre in size. Suction or pressure assisted cups are usually larger and mounted below the gun for better balance. Sprayguns can be fed directly with paint under pressure from larger containers.

Figure 129 Don't forget to check HVLP air cap pressure

Spraygun set-up

Before any spraying can begin the spraygun is set up by selecting the correct combination of needle, nozzle and air cap. Many makers supply the nozzle and needle as a matched set. Sometimes the air cap is also matched to the nozzle set-up, although there are occasions when it is better to vary the size.

The size of the nozzle and the matching needle is selected for the type of paint and its viscosity. Paint makers list either a set-up code or give the nozzle diameter in millimetres. The spraygun information usually gives both. The air cap must be matched to this assembly to provide enough air to carry the paint. Large jobs such as complete resprays demand a high rate of paint transfer to ensure an absence of banding or dryness. It may be better to use a larger nozzle for this purpose, provided that there is an adequate air supply to support it. This will be revealed by a maintained pressure on the air micrometer. Full information on air supply will be found in Chapter 6.

Conventional sprayguns demand pressures up to 5.5 bar (80 lbf/in^2) to work correctly. HVLP can operate at pressures between 1.4 bar and 3.4 bar (20 lbf/in^2 and 50 lbf/in^2), although at least the same quantity of air, or even more, may be needed. Similar nozzle and needle sizes are given for HVLP. Air supply at the inlet can be measured with an air micrometer in exactly the same way as for conventional sprayguns. What is different is the air pressure reading for the air cap. The air pressure within the cap must not exceed 0.7 bar (10 lbf/in^2) and is set to varying figures below this reading to suit the material being sprayed. Spraygun instructions invariably give an inlet pressure that should provide the correct figure at the cap. It is good practice to measure the outlet pressure at the cap occasionally (see Fig. 129). For this purpose air caps are available with one or two gauges. The two gauge cap will ensure compliance with the regulations under all circumstances.

The combination of mixing and thinning data, nozzle and aircap sizes, and inlet air pressure normally provides a spray pattern which only needs minor adjustment. With so many variables there will be times when the set-up may need to be found by experimentation. This will certainly be the case when no figures are available. Figure 130 is a modified extract from ICI, 'The Refinishers Handbook', which will enable you to do just that.

Faults from too high an air pressure setting are:

- Cobwebbing,
- Dry spray,
- Low gloss.

Faults due to the pressure being set too low include

- Floating,
- Popping,
- Orange peel,
- Pinholing,
- Runs and sags.

Examples of these defects can be found in the coloured plates section.

The pattern of spray (see Fig. 131) is fundamental to good spraying and, if not already tested, must be checked before commencing every job. The ideal shape is an oval with almost straight sides. There are four basic pattern variations that will indicate a fault (see Fig. 131):

1. Heavy side pattern shown by a crescent moon shape to left or right.
2. Top or bottom heavy pattern shown by a pear shape pointing up or down.
3. Split pattern in an hour glass shape.
4. Heavy centre pattern shown by a pointed oval.

Fortunately it is fairly easy to locate the exact problem. Unbalanced shapes, 1 and 2, are almost certainly in either the air cap or in the paint nozzle. The even, yet incorrect shapes 3 and 4 are the result of maladjustment or wrong paint viscosity. Here is how you can check them:

- The heavy side pattern, a crescent moon shape, should be checked by turning the air cap around by half a turn. If the shape goes to the other side, the opening in the horn of the air cap is blocked. If the shape does not move there is a defect in the nozzle.

How to Determine Optimum Spraying Pressure Accurately and Quickly

This simple check will enable the operator to determine the optimum spraying pressure for any paint at any reasonable viscosity, with any spraygun.

NOTE:
Spraying pressures either higher or lower than the optimum may be necessary in certain special circumstances: e.g. obtaining particular metallic effects, working with exceptionally fast or slow thinner, or extremes of climate.

1.   Attach sheet of paper to wall.

2.   Adjust air pressure to approximately 30 psi (2 bars).

3.   Adjust spreader and fluid needle controls on spraygun to fully open position.

4.   Hold gun a set distance from paper and spray for 2–3 seconds keeping the gun stationary and perpendicular to the surface. A quick easy way to establish the distance – which should be in the region of 6–9″ (15–22.5cm), is to use the distance between the thumb tip and little finger tip of the outspread hand.

5.   Raise pressure by 5 psi (0.3 bars) and repeat.

6.   Continue to repeat, raising the pressure by 5 psi (0.3bars) each time, until the pressure which gives the maximum spray pattern size is established.

7.   Attach a clean sheet of paper to the wall. Keeping the gun perpendicular, and the same distance from the paper (as in 4.), make a very fast pass across the paper holding the trigger right back (gun fully open).

NOTE:
The gun *must* pass fast enough for the paint particles to fall *separately* upon the paper.

8.   Raise the pressure by 5 psi (0.3 bars) and repeat. Compare the paint particle size.

9.   Continue to repeat, raising the pressure by 5 psi (0.3 bars) each time, as long as necessary to determine the pressure which gives the finest, or at least adequate, atomization.

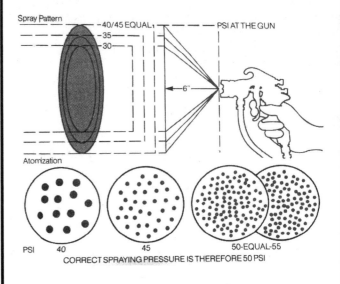

Figure 130   Setting up a spraygun
Source: J. J. Beaumont in association with B. D. Fitton, 'ICI Autocolor – The Refinishers Handbook' (ICI Autocolor, Wexham Road, Slough, Berkshire (1983, p. 32).

■   Heavy top or bottom, pear shaped pattern is checked as above.

■   The pointed oval of the heavy centre pattern indicates that the set-up is not right, the settings are incorrect or that the paint is too viscous for the set-up. If you know that the viscosity is correct, re-check the paint maker's data sheet for the set-up details. If those details are correct, adjust the air pressure or the fluid flow settings.

■   Split spray patterns of an hour glass shape are due to incorrect spraygun control adjustment. Re-balance the air and fluid settings.

When cleaning is needed, the air cap and nozzle must be cleaned with correctly shaped needles and brushes from the spraygun maker. Do remember that the openings drilled in the air cap and nozzle and the tapered end of the needle are to exact dimensions and to a high-quality of surface finish. On no account use any other implement to clean out the openings. If there is any suspicion that the needle is bent then it must be straightened or the nozzle and needle set replaced. The outlet of the nozzle should also be of an even, smooth shape without any damage. Again, if this is damaged the nozzle and needle set must be changed.

Spraying techniques

The general rule is to hold the spraygun at a right angle to the face of the panel (although there are times when some deviation is permissible or even desirable). Normally this is essential for an even coating of paint without unwanted side effects. This position must not change throughout the stroke of the spraygun from one side to the other. The body, arms and the wrist must all be free to maintain this movement. In this way an even coating of paint can be laid onto the panel (see Fig. 132).

An even coating also depends on the speed of passing the gun across the panel, on the type and viscosity of the material and on the gun set-up. The settings are largely personal to you and will be matched to your normal speed of gun movement. The aim is to obtain a good covering without an excess that would cause runs or sags. On most car work, the length of each stroke is governed by the width of the panel. Where you have a choice, choose a length of about 90 cm (36 in) that you feel happy with.

The quality of the coating is dependent too on the distance of the nozzle from the surface of the panel (see Fig. 133). Conventional sprayguns allow a little tolerance with a

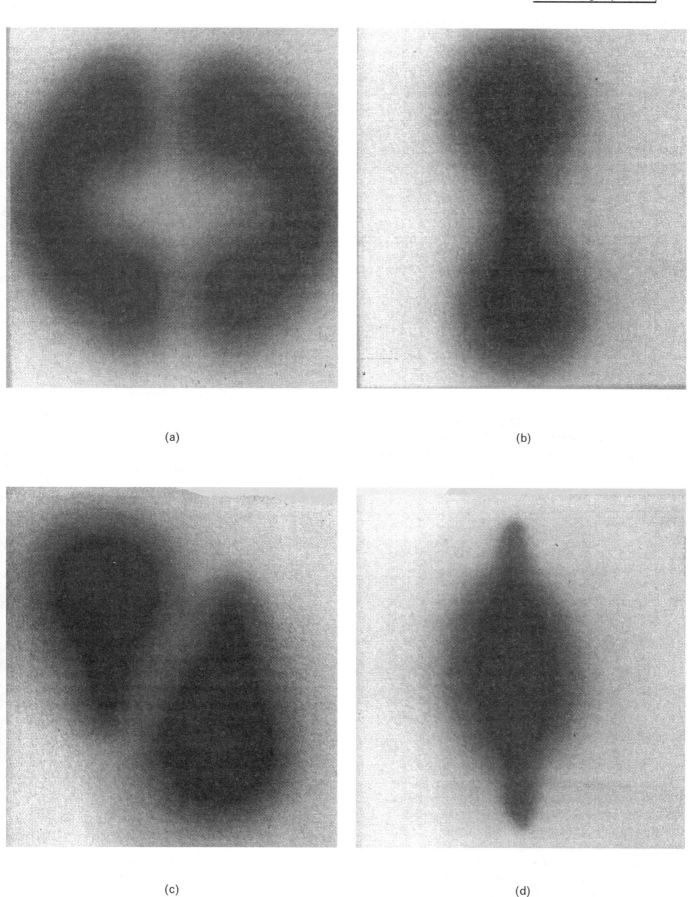

(a)

(b)

(c)

(d)

Figure 131 Defective spray patterns:
    (a)   Heavy one side indicates a horn hole blockage or fluid tip obstruction. Rotate air cap to check the cause
    (b)   Improperly balanced air and paint flow
    (c)   Horn hole or fluid tip obstruction at top or bottom
    (d)   Heavy centre pattern is caused by incorrect spreader valve adjustment, paint too thick, air pressure too low or incorrect size of fluid tip

distance of between 15 cm (6 in) to 20 cm (8 in). HVLP guns are a little more demanding, most makers of paint or sprayguns recommending between 15 cm (6 in) and 18 cm (7 in). A distance of 20 cm (8 in) will usually give a dry spray.

The next important consideration is trigger operation. The trigger should be released just before the end of the movement and pulled again just after starting the return pass. The easing of sprayed paint at each end will provide a natural fade-out that will also prevent excessive build-up on overlaps.

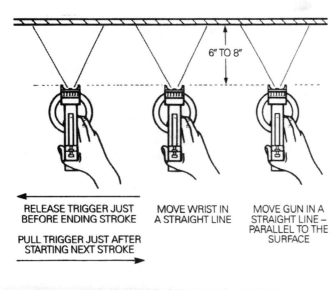

RELEASE TRIGGER JUST BEFORE ENDING STROKE

PULL TRIGGER JUST AFTER STARTING NEXT STROKE

MOVE WRIST IN A STRAIGHT LINE

MOVE GUN IN A STRAIGHT LINE – PARALLEL TO THE SURFACE

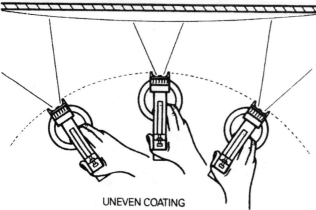

UNEVEN COATING

Figure 132 Spraygun techniques

Begin painting by spraying any edges and openings. Moving onto the panel, paint is laid on in stripes which should normally overlap by half the width of the stripe. This is achieved by aiming the gun at the edge of the previous pass. Where vertical sections are being painted, such as the side of a van (see Fig. 134), start each section at the top and overlap the previous one by about 10 cm (4 -F7in). When spraying a good part or all of a vehicle, it is important to have a clear idea of how you intend to move around the job. A car respray, for example, may be tackled by painting the roof, dropping down to the front door, then the front wing and across the bonnet to the other side (see Fig. 135). Continue along the side, over the boot or tailgate and return to the start point. Start and end at natural break lines such as a door pillar.

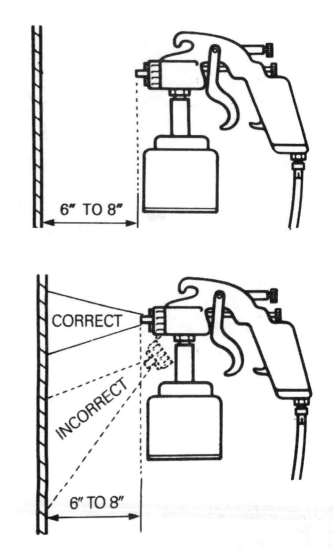

Figure 133 Gun distances are critical, especially with HVLP

Direct gloss or solid colour coats are the easiest of topcoats to apply. Conventional, non-high solids paints may need two or even three coats for coverage. Alternatively, there are now high solids materials that will cover in one pass to depths of up to 75 microns. The process then will vary greatly and depend upon the specific instructions for the paint that you are using and the spraygun. Multiple coats are invariably applied 'wet-on-wet', always allowing the necessary 'flash-off' time between coats. Times range from just a few minutes to 15 minutes or more. Flashing-off is particularly important as solvents may be trapped if a new coat is applied too soon.

Paint coats that are applied too heavily or second coats applied too soon may give rise to a variety of defects. Here are some of them:

- *Floating, flooding, mottling or shadowing.* This is a variation of colour with lighter and darker areas caused by the pigments moving in the fluid paint before it has set. The same defect in metallics is caused in a different way and this is explained in the section on Colour earlier in this chapter.
- *Wrinkling, crinkling, puckering, rivelling and shrivelling* are all names for the irregular ridges and furrows that can occur with synthetic paints that are applied too thickly.

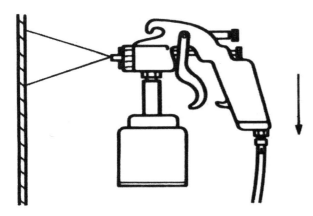

WORK DOWN PANEL APPLYING CROSS COATS WITH GUN HELD AT RIGHT ANGLES

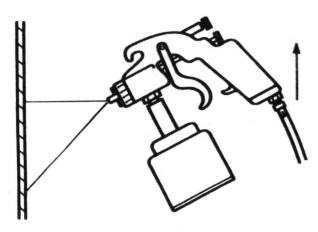

THEN WORK BACK UP THE PANEL APPLYING CROSS COATS WITH THE GUN HELD AT 45° TO THE SURFACE

Figure 134 A technique for working up and down a panel, such as a van side

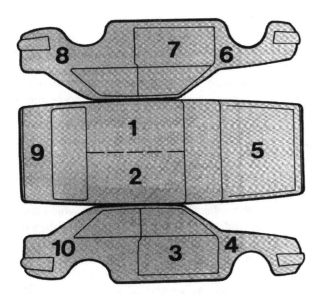

Figure 135 A suggested sequence for a complete respray
Source: Reproduced by kind permission from 'ICI Autocolor – The Refinishers Handbook'.

This is caused by the top layer of paint drying before those below.
- *Boiling or solvent popping* is caused by escaping solvent trapped by overcoating too soon.

There are other products marketed by the paint companies which permit easier work, provide coatings for plastics and special effects for particular models. One example is the additive that may be used for the paint instead of an isocyanate hardener to simplify painting the interior, or the door shuts.

Basecoats are mainly metallic finishes, although the application of clear lacquer to dark solid colours is an established procedure. The clear lacquer provides the gloss to metallic finishes and enhances the gloss on dark solid colours while at the same time minimising the effects of scratching.

The various paint systems vary in the use of solid colour. Where solid colours are used as a basecoat, a stabiliser is sometimes added to the base material instead of a hardener. In such cases the paint should be applied in modest coats to avoid slow drying. The time available in which overcoating with the clear lacquer is done varies with the product. Always follow the instructions on the maker's data sheet. Where it is necessary and permissible to allow the basecoat to cure for a long period, such as overnight, it may be necessary to reactivate it with another basecoat before applying the clear lacquer.

Metallic basecoats are either a single colour or may be of the pearlescent type. Normal basecoats consist of a tinted translucent paint film containing many tiny platelets of aluminium. In conventional materials the platelets are of irregular shape and often become broken even further during application, reducing the overall light reflection from the coating and giving an impression of dullness. A more recent development has been the introduction of 'lenticular' tinters which contain lens shaped platelets. These have much greater reflective effect giving a brighter, cleaner colour.

Whichever type of metallic you are spraying, the mixing, gun setting and application will determine the colour shade, face and side tones of the finish. To match existing paintwork it is necessary to determine not only the colour shade but also the type of metal particle used and how it is laying inside the basecoat. This is where a painter's own colour library can be so useful and add to the information provided by the paint maker. Start with the basic formulation and spray a test card with a coat of clear lacquer. The lacquer plays a large part in the appearance of the finish and it must also be added to all spray-out cards when it is used to finish the car. The colour can be compared, together with the side tone, from which three decisions can be made:

- Colour correction using tinters. Before making a decision spray more cards using different gun settings or distance.
- The brightness of the reflected light will determine if the metal particles are a matching type. If not, it may be necessary to try a different mix.
- The variation between face and side tone will indicate how the metal particles are laying in the coating. Try applying a deeper or shallower coat. Remember that some tinters may have more effect on one tone than on the other.

Matching metallics is not usually as difficult as it sounds because the very factors that can cause you so much difficulty also come to your aid. The fact that face and side tone are so often different means that panels can be painted up to a break line or edge. Provided that there is a clear change in the angle of the panel, a good fade-out at that point means that there is no direct comparison of the colour, as it is not possible to see both your new paint and the original at one and the same time. Other information and details of some of the main considerations when mixing metallics will be found earlier in this chapter, on p. 98–100.

Conventional basecoats are often applied in three stages; two full wetcoats followed by a thinner finishing coat. Some makers recommend the second coat being applied with the gun held a little further out from the panel. Take care to follow the flash-off instructions. Do not forget the colour card for your library.

The basecoat is allowed to flash-off to a matt finish before the clear lacquer is applied. Two or three coats, depending on the system being used, are sprayed wet-on-wet. Flash-off times of a few minutes must be allowed between coats. The flash-off times are very important. Too much time will result in poor adhesion between the clear lacquer and the basecoat. A shorter period may result in the defect variously described as clouding, mottling, flooding, floating and shadowing. This is a movement of both the colour pigments and the metal particles in the basecoat caused by the basecoat dissolving and the particles floating into different positions.

The ultimate in the art of painting is the application of pearlescent or pearl effect coatings. These finishes comprise three main elements: a base or background colour, a translucent coating and the familiar clear lacquer. These may be applied as three separate coatings or the base colour may be combined with the translucent coat. A two coat system is handled exactly like the normal basecoat and clear, the only exception being the considerations of matching the pearl effect. The special effect of both these paints is created by adding to either the translucent coating, or the combined basecoat, particles of mica instead of aluminium.

Refinishing or totally new application demands the highest standards of workmanship. A successful result with a pearlescent finish, within the normal limitations of paint depth, is a sign of a true craftsman. The author has known of these finishes being applied coat upon coat until the painter was satisfied. In one case this resulted in a paint depth of over 800 microns! The secret with these finishes is to have a perfect foundation; the slightest blemish will show through the other coatings. The other factor is that these paints may change in effect with successive coats. If blemishes show through, or an excess of paint is applied, the only satisfactory rectification is to go back and prepare the surface correctly. This should be followed by a gradual build-up of the paint film, carefully checking each stage.

Most pearlescent finishes are based on a white or other light shade of basecoat. In three coat work, two to four coats of the translucent coat are applied, depending upon the make of paint. A film build of between 15 to 30 microns is desirable. This is followed by the usual clear lacquer, applied to a surface which has flashed-off to matt. Some pearlescent basecoats contain large particles and the spraygun paint filter (if

fitted) may need to be removed. This should only be done in exceptional circumstances as directed by the paint maker.

Drying and curing

When spraying is complete most finishes benefit from curing or drying by the use of baking or low stoving. This can be done by using hot air in the oven or combined booth and oven, or by infra-red heating elements. Hot air drying or curing is generally carried out at temperatures between 60° Celsius and 80° Celsius. Generally, paints are cured or dried at the lower temperatures. Only use a higher setting if the paint data sheet indicates that this is acceptable. Most finishes will air dry without heat but longer times give a greater risk of contamination and a much longer time before they are sufficiently hard to withstand contact (see Fig. 136).

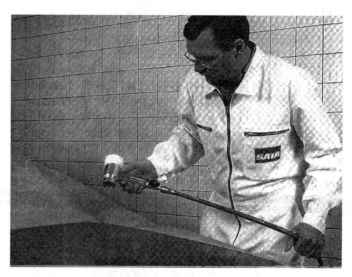

Figure 136 The drying of water based paints can be delayed because the air in contact with the surface is already saturated; moisture removal from the finish is accelerated by increasing the airflow – this hand held unit is one method

All paint curing ovens, which are almost always combination booth-ovens, are dependent on correct setting-up and maintenance. You will find details of these subjects under Equipment and Tooling earlier in this section. Complete sealing of the booth and a slightly negative pressure are absolute essentials to comply with regulations and protect people's health.

Drying with infra-red is now commonplace in the industry. There are a number of points to be considered if the use of this system is to be successful. The radiant elements of the drier unit heat the substrate of those areas exposed to the rays. There are two implications. Parts of the car which are hidden from direct radiation, such as door shuts, will cure by the conduction of heat through the substrate but may take a little longer. As the substrate below the paint is heated, heating is from the inside outwards. It is possible that substrate temperatures may rise beyond 80° Celsius and could exceed 100° Celsius on some occasions. Any temperature sensitive components attached to the substrate, either inside or out, may need to be removed or protected.

Infra-red equipment is capable of drying all types of materials, including water borne. They have the great advantage of much reduced energy consumption and fast curing times, even complete resprays taking as little as 15 minutes. Many types and sizes of equipment are available for almost any process needing drying or curing. Setting the infra-red heater for use may well involve colour correction of the drying time, particularly on units with automatic controls (see Fig. 137).

Care must be taken to protect unshielded radiation units when they are not in use. Dust or overspray can reduce their efficiency dramatically. The usual care of electrical equipment and cables applies to these units.

Post-cure care

All paint films are susceptible to damage when freshly cured, either because they are not fully hardened or because they do not yet have a protective coating to guard against the elements and other damaging agents. In particular, freshly cured and unwaxed paintwork should not be subjected to rain followed by sunshine. This will almost certainly result in rings of contamination on the surface which may prove difficult to remove. Full details of this and many other sources of damage are given in Chapter 8.

(a)

(b)

Figure 137 Infra-red drying arches are profitable for high throughput paintshops; the arch in (b) moves between two paint booths to increase productivity

Identifying and Rectifying Finish Defects

The three sections of this chapter have much in common with topics dealt with previously. This chapter is confined to the new elements of paint defect identification, prevention and rectification, where this has not already been covered.

Health and safety at work and PPE

There is little risk attached to the use of most materials and equipment that will be needed for paint defect identification and rectification. Nonetheless, it is prudent to protect yourself from the long-term effects of mild chemicals, particularly your hands.

Rectification is either by surface treatments or refinishing, when all the usual protective measures must be employed. Surface treatments mostly involve the use of flatting or polishing materials, which normally only call for the use of disposable gloves. However the makers' instructions should always be checked and followed.

When oxalic acid, which is highly poisonous, is used, full protection including footwear, must always be worn and the instructions for use carefully followed.

Other matters of health and safety are dealt with in Chapters 6 and 7, to which you should refer.

Housekeeping and Waste Disposal

Most of the residue from paint inspection will be polishing cloths and wipes. These may be put into normal waste. Solvent laden cloths or wipes are treated as paint materials and put into water damped, closed containers. All painting materials are handled in the usual way, of course.

Oxalic acid will be used diluted 1:10 with water and then rinsed away with copious amounts of water.

Some rubbing compounds can create a very slippery surface if they are dropped onto the floor. If this happens the area should be kept clear and the material cleaned up as soon as possible.

Sources of Information

Refinishing calls for the use of all the normal refinishing materials and possibly some special sealers. The usual paint makers' instructions should be followed.

The trade press will also cover this subject occasionally. This may be the only way in which you will be able to keep up to date with information from sources that are not available to your bodyshop.

Identifying defects

Equipment and tooling

No close inspection can take place until the paintwork is clean and so a power washer or a clean bucket, sponges and leathers are essential. Notes should be taken down on a form suitable for filing as a permanent record, on a notebook computer or entered directly into a computer terminal.

A good magnifying glass and, ideally, an illuminated viewer are necessary for inspecting paint (see Fig. 138). Many conditions are so small that accurate identification of the defect is not possible without these items.

Hazard	Protection
Washing down paintwork	Non-hazardous wet wear and rubber boots Gloves to protect the hands from the drastic effects of cleaning agents
Cleaning, polishing and cutting back paintwork	Disposable gloves unless otherwise stated
Solvent tests to determine the type of finish	Solvent proof gloves Cartridge mask to prevent inhalation of fumes
Using oxalic acid to identify and remove ferrous particle impregnation Using paint stripper to remove paint coatings	Acid proof gloves, face visor, fume mask, acid proof apron and rubber boots

Job	Source of information
Cleaning paintwork	Product instructions Product makers' information Suppliers' leaflets Paint makers' data sheets
Using equipment	Equipment makers' instructions Paint makers' information Vehicle makers' information
Identifying paintwork defects	Vehicle makers' information Paint makers' information

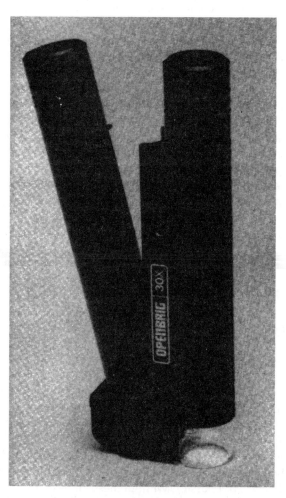

Figure 138 A lightscope giving an illuminated area at 30X magnification, which is ideal for paintwork examination

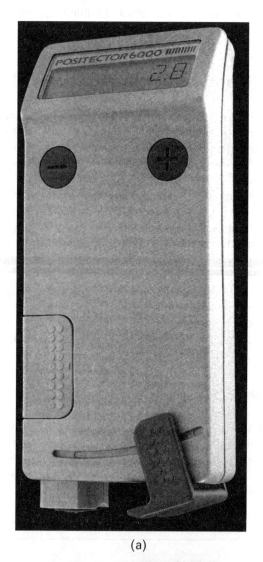

(a)

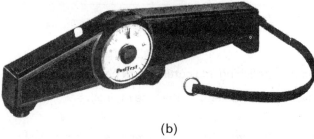

(b)

Figure 139 Measuring paint finish depth is a fundamental part of defect diagnosis – both (a) electronic and (b) magnetic gauges are available, and magnetic gauges must be re-calibrated by the maker periodically

The depth of paint is of crucial importance. A depth gauge indicates to a painter how the preparation for painting should be done (see Fig. 139). Identification of defects and general inspections of paint condition also demand an accurate measurement of the depth of the coatings on the vehicle. Lever type magnetic gauges are quite adequate for all normal work on ferrous substrates. One disadvantage of these is that they must be sent away for calibration periodically. Electronic gauges are better because they are either self-calibrating or can easily be calibrated on the spot. Some also have the advantage that they can measure paint depth on non-magnetic metals such as aluminium.

Although not usually considered necessary for normal refinishing, other gauges and meters are available for such things as the measurement of orange peel and the checking of colour.

An electric or pneumatic polisher with disc pads and polishing heads will be needed for surface rectification work. A special 'run cutter' or a piece of body file will be found invaluable for cutting back the bulk of paint runs. Flatting and polishing materials will be needed for rectifying surface defects. Fine abrasives and a small de-nibbing block are essential for remedying small inclusions.

Refinishing equipment and tooling is identical to that used for normal painting and is dealt with in Chapters 6 and 7.

Identifying paint finish defects

Paint defects can be divided into three main groups (see plates 1–29): those that occurred in the factory, those caused through use of the vehicle and some perhaps as a result of refinishing. Defects occurring through use of the vehicle can be influenced by the user, such as stone chipping, while those from unavoidable causes, such as acid rain, can affect every vehicle exposed to the elements.

The standard of most factory finishes is very high, especially when the continuous throughput of the production line is considered. Indeed, cleanliness is of such a high order that most bodyshops find it difficult to achieve 'from-the-gun' finishes as good as the factory when repainting a panel under warranty! Here are some examples of what you may expect:

- Lack of adhesion (see plate 15), shown by a readiness to chip badly from moderate stone chip impact. This is usually due to overbaking of one of the coats due to the production line slowing down or stopping while the body was in the stoving oven.
- Soft finish is occasionally found. Here the impact damage from stone chips will be a wider topped crater. The surface will also scratch very easily.
- Filler coat 'grinning through' or lack of cover (see plate 12). Often occurring around the lower parts of the car or on inside areas. The cause may be a reduction in the paint supply to the robot or a human sprayer being a little too quick in moving the spraygun.
- Clouding, mottling or uneven streaks in metallics sometimes occur with factory paintwork due to production line problems (see plates 9 and 29). Where vehicles are painted by human hand using more than one painter, differences between their work may be too obvious.
- Dirt inclusions do sometimes occur, despite multiple washings in demineralised water and drying with hot air. It is the same problem that you have in your bodyshop. The only difference is that the factory production line paints hundreds of bodies in a day!
- Substrate contamination can be seen occasionally (see plate 27). This is usually of a small, localised nature such as a smear of grease on a panel. Such marks are most likely to occur through a drip of lubricant from a mechanism on the line. They appear as blisters, sometimes as much as eighteen months after production. They

often have a distinctive spider's web appearance, most noticeable after they are flatted.
- Orange peel is rarely excessive these days (see plate 19). Where it does occur to excess there is every chance that the paint or the application was not correct. This 'defect' is very much a matter of opinion as mass produced vehicles are made and finished to an acceptable standard, not perfection. Indeed, amongst the equipment made for production line use are orange peel gauges, to permit a standard to be set and measured.

Before any work is carried out on correcting an original finish, particularly under the vehicle maker's warranty, your bodyshop will have made sure that it has been approved (it will be paid for!) and that you are able to improve the finish.

During everyday use paint finishes on vehicles are subjected to one of the harshest environments that it is possible to find. They are frozen in winter, baked in the sun, washed in acid rain, often with industrial additives, struck by stone particles travelling at up to 160 miles an hour, attacked by car washes and abraded by shopping baskets and hedgerows.

It is not surprising that they suffer damage, often due to too little care on the part of the user!

Most vehicle makers provide some advice on how users should care for bodywork and, in some instances, it is very good. It is all about preventing, or at least limiting, the damage caused through use. Your customers rarely know the correct way to care for their vehicles. At the end of this section there is some general guidance which you should study and pass on to your customers when opportunity offers. The list that follows gives many of the finish defects that can occur in use. First those that can be minimised or even avoided by the driver:

- *Stone chipping* affects almost every vehicle (see plate 25). Driving on loose, unmade surfaces can cause a lot of damage to the lower parts and underpan. Driving too close to the vehicle in front, particularly at high speed on motorways, can also cause a lot of damage to frontal areas of the body. The damage caused may be quite different to that caused by loose surfaces. Slower speed, heavier stone impacts will create larger craters easily visible to the naked eye. The damage from motorway use is often only seen initially through a magnifying glass. Here, tiny, very sharp stone chips can strike the paintwork at high speed and create deep, almost invisible craters. Many of them will cut through to the substrate, allowing corrosion to begin. This defect may only be noticed by the user when corrosion bubbles first appear. Viewed through a glass or an illuminated viewer, the original opening in the paint film will be clearly visible.
- Scratching caused by an *automatic car wash* is also usually invisible, becoming noticeable through a loss of gloss (see plate 10). Where the paint is unusually soft, scratching may be seen. If the scratches cannot be polished out the surface is too soft.
- *Black or dark stains* in the paint, particularly on white or other light colours. These are frequently caused by incinerated particles of ash or soot falling onto the paintwork (see plates 8 and 14). Industrial premises, waste disposal facilities and even hospitals may give rise

to this defect. Coal burning power stations are particularly bad unless their chimneys are fitted with filters. Paint staining is caused by the acidic content in the ash.

■ *Rusty marks or a sandpaper effect* on the surface of the paint are caused by parking near heavy industrial premises such as foundries, or close to busy railway stations. It is a widespread problem in the South East of England, where hundreds of thousands of cars are parked everyday at or near busy stations. It is also a problem for garages and bodyshops in this area. The defect is caused by the hot metal particles worn from the train brake shoes during braking being blown onto the nearby cars. The particles attach themselves to the surface of the paint film. Before rusting the sandpaper effect can be felt by passing the fingers lightly across the upper surfaces.

■ Small drops of an *amber fluid which has set on the paint surface* (see plate 5). This is most likely to be resin, which has dropped from the foliage of a tree onto the car parked below. The resin will react with the binder in the paint film under the influence of heat, amalgamate with it and be very difficult to remove.

■ *Yellow rings about 1 millimetre across*, or the same size ring of topcoat eaten out of the paint film, are also caused by parking under trees (see plate 1). This defect is a result of the honeydew produced by aphids dropping onto the vehicles below. Heat and a high level of humidity will enable the deposit to eat into the paint film.

There are also more general hazards to paintwork that can cause damage wherever they are in the open. Here are some of them:

■ *Bird droppings* (see plate 4). Pigeons and gulls that feed on waste tips are the worst offenders as far as damage to the paintwork is concerned. The content can be highly corrosive and, if left on the paint in times of high heat and humidity, can eat into the paint film.

■ *Yellowish brown circles about 3 millimetres across* will probably be the droppings of unhealthy bees (see plate 2). They contain insecticide residues which can again, during a period of heat and high humidity, cause permanent damage.

■ *Small, circular marks on, or in, the paint film* (see plate 3). Close examination may reveal the shape of an insect, such as a wing. During hot, stormy weather in the summer, some flies and mosquitoes mistake the shimmering surface of a dark coloured car for that of a pond. They land on the paint, stick to the hot surface and die there. Their bodies are highly acidic and, if left, can eat through the topcoat in a short space of time.

Many of these defects are not easily determined without close examination. It can also be very helpful to know how and where the vehicle is used.

Refinishing defects can be of many different types, and can be caused by the refinishing materials, the condition of the substrate or the methods of application and curing. The age of the materials may also affect the finish. It must also be remembered that where supposedly unrepaired vehicles are concerned, in fact refinishing may have taken place in the factory, during transportation or in the supplying dealership. At such times the process will almost certainly have been the same or similar to that which you use in your daily work. Measurement of paint depth is an almost certain indicator. These defects have been grouped together by similarity of appearance to assist in identification.

The first group are those which are predominantly on the surface or in the top of the paint film:

■ *Inclusions of dirt or fibres* are one of the most common defects (see Fig. 140). Many are caused by dirt from outside or even inside the booth, by fibres from clothing, or by failing to carefully clean the work before applying the paint (see plate 26).

Figure 140 One fundamental reason for inclusions is particles in the paint; a paint strainer should always be used unless the maker instructs otherwise

■ *Gritty feel and appearance of the paint; specks of matter visible on the surface.* This may be due literally to sand in the paint which settled on the surface during painting! However, it is more likely to be caused by a fault in the paint due to poor storage, poor mixing machine operation or the product past the 'use-by' date. Most paints have a shelf life no greater than 2 years. The batch number can be used to verify the date of manufacture from the makers.

■ Small craters in the topcoat where the paint has rolled back from the surface below is an almost certain sign of *silicon contamination*. Variously described as cissing or fish-eye, it can be caused by contaminated air but this is more likely to affect every job (see plate 17). Small and isolated or large individual craters or roughness may be caused by surface contamination from another cause. Finger prints, drops of oil, cleaning agents and moisture may all be responsible.

■ Tiny, rounded blisters of even size, often with a tiny opening, covering the paintwork following refinishing. Usually known as *solvent boil or popping*, due to either air or solvent trapped inside the paint film during drying (see plate 16).

- Similar blisters of varying size and shape, although not usually burst, indicate the presence of either *solvent or water trapped in the film*. They may come and go depending upon the weather; rising when hot and disappearing when cold.
- *Projections and or dark spots all over metallic basecoat* suggest a basecoat mixed and applied after the 'use-by' date (see plate 24). An incorrectly sealed canister leading to exposure to the air or humid conditions may also be responsible.
- *Wrinkling* of the paint surface is very like the patterns created on the sand of the seashore as the tide goes out (see plate 22). This is caused by either too much paint applied at one time or the surface film drying before the paint below.
- *Regular grooving predominantly in one direction* is flatting marks on the coat below which have not been sanded away (see plate 23).
- *Sags and runs* are too well known to need description (see plate 18). Caused by an excessive amount of paint in that area and for the conditions.
- *Mottling* may take place if water collects on a paint film which is not totally cured, particularly rain which falls after a period of dry weather (see plate 28).

The second group are those which can be seen in the paint film or changing the colour:

- *Varying shades of colour* in either solid colours or metallics are caused by uneven application of the paint film.
- Very similar is the shading effect caused by *floating of the pigments* in paint film. Incorrect thinners and application techniques can both be responsible.
- *Cracking or crazing* of the paint film can be caused by a fault with the material or by excessively deep paint coating depth (see plate 21). Mixing faults, incorrect materials, sub-surface defects, together with a paint depth greater than 250 microns may all be causes. The condition may also be called 'checking' or 'crow foot'.
- *Lack of gloss* with some texturing is evidence of a number of faults (see plate 11) including topcoat applied before the filler had completely cured, too heavy application of topcoat, use of incorrect materials and poor finish to a coarse filler (see plates 7 and 10).

Inspection preparation

Many of the tell-tale signs of a paint defect are very small. In order to see them three conditions must be met: the paint must be clean, the lighting must be good (where colour is being checked it must be natural or a specified light) and you need adequate equipment. Where the inspection is carried out as a result of a complaint, the customer may well be present. If either a vehicle manufacturer or a bodyshop is not liable for the cost of refinishing, the major problem is convincing the customer. In such circumstances you must not only know what you are doing, but look as if you do!

The cheapest and best aid for that is a clipboard and an inspection form. The form may be a manufacturer's official document or one that is home-made. The important point is that you are making a record of your findings which is going to be kept for future reference. It is sometimes appropriate to give the customer a copy.

An adequate examination cannot be done without a good magnifying glass or, better still, an illuminated viewer. In many cases you will need one of them to see the defect and to show the customer. A paint depth gauge is essential, particularly where there is a suspicion of previous refinishing. Reference material such as manuals or fault finding guides should also be on hand to aid identification and, if appropriate, to convince the customer.

Before you begin, the paintwork itself must be free of road dirt. If a wash is necessary, take note beforehand of the basic condition, particularly those things which show how the customer has cared for the vehicle. This may be of critical importance where refinish under a vehicle manufacturer's warranty is being claimed. It is also useful to know if the customer keeps the vehicle waxed. This can be checked by pouring water onto a level panel. If the water collects into globules there is a good coat of wax; if the water spreads there is little or no wax on the paint.

Inspection procedure

Plainly, checking your own work only calls for examination of the finish, and perhaps nearby paintwork that has not been refinished for comparison. You need only carry out the suggested checks and tests that are relevant. Here now is the full procedure for a vehicle that you have never seen before with a customer present.

Once the vehicle is clean the inspection should be carried out carefully and systematically using the following procedure or a similar one. Involve the customer, asking relevant questions and explaining what you are doing in a friendly but efficient manner:

- Place the vehicle in a good northerly light without sharp contrasts nearby. Remember that where a complaint about colour is concerned, some paint makers specify the precise lighting for comparison.
- Note the vehicle identification number.
- Walk around the vehicle making a note of the general condition and plainly visible defects or special features.
- Check each major panel and the front wings and rear quarter panels for paint depth. Greater depth than usual will indicate refinish although this may have been OE (original equipment) rectification.

Then follows the detailed examination of the area of complaint. Clear information is needed on several aspects if future difficulties are to be avoided. If a manufacturer's paint or body warranty is being invoked, they will need to know that the finish is original and that the problem is a manufacturing defect. Generally the customer must also have complied with any conditions of care, which can range from washing and waxing the vehicle regularly to body inspections by a dealer at set intervals. Prompt repair of damage is usually a condition, too. If the complaint is about work carried out by your firm, it is essential to confirm that the job was done there. Detailed examination can now begin:

- Examine each panel which is the subject of the complaint, not forgetting to keep the customer informed and involved; useful information is often given in conversation. Carefully log all defects, including stone chipping and if it has been touched-in.
- Check surrounding panels or others that you would expect to be affected in a similar way. For example, if the customer complains of roughness on the bonnet you repainted three months ago and it is train brake or grinder dust as you suspect, it will be on other panels too. If its only all over the bonnet it is almost certainly down to you.

Making a correct diagnosis of the defect depends upon the inspection. Correctly identifying the often tiny evidence by magnification and comparison with photographs is the only way to become proficient at this work. Do remember that an incorrect diagnosis can lead to great expense in time, effort and materials with no guarantee of customer satisfaction at the end.

Interpretation of results

Once the defect has been clearly identified the cause is often known or is of no concern. For example, contamination during manufacture which has resulted in a blister. All that is now needed is to strip back the paintwork in that area and refinish the panel with a local repair to the vehicle maker's specification. A defect arising from a failing during refinishing will involve stripping back to the substrate or sealing underlying coats to provide a good finish. Interpretation of the cause of damage needs a little more thought where the customer's own subjective views clash with the reactions of others. Stone chipping, particularly on relatively young cars, is nearly always a disputed issue. The customer often blames the vehicle maker or the bodyshop when there is nothing wrong with the paint; they drive without exercising proper care or use the vehicle in circumstances where the paint cannot avoid being damaged. Off-road vehicles used for that sport are a case in point. How then, are you able to decide the difference between a defective finish and just the results of hard use?

There are unlikely to be any vehicles that do not sustain stone chip damage as most of the impacts of tiny grit particles are so small that they cannot be seen with the naked eye. In the normal course of events if the vehicle is waxed or washed sometimes with a wax shampoo, the tiny openings are filled with wax and are never seen. However, most vehicles suffer the occasional heavier impact which breaks away enough paint for it to be noticeable. The fact that there is one or perhaps many of these stone chips does not automatically show that the paint is defective.

There are two aspects to paint and how it withstands impacts; they are called cohesion and adhesion:

- Cohesion is the bonding of the paint particles together.
- Adhesion is the linking of one coat to another.

A weakness in both these aspects will result in more damage than usual to the paint film from the impact of stone chips.

A lack of cohesion is weakness within the structure of the topcoat itself. The paint is gouged out without necessarily penetrating right through to the underlying filler coat. Looked at under strong magnification, the opening is like a crater, often with different colour tints visible. There are likely to be many of these impact craters visible on front facing panels and the front of the roof if the whole of the vehicle has paint in this condition. If the damage follows repainting, only those panels which have been refinished will be affected, of course.

Adhesion weakness is the more usual problem, through the topcoat not having a key onto the filler coat. On OE finishes this is often due to the filler coat being overbaked, resulting in a smooth, glossy surface to which the topcoat cannot adhere. Poor adhesion is shown by the loss of flakes of paint, often semi-circular in shape, which expose the filler coat. Under a magnifying glass they can be seen to have cliff-like edges. Sometimes adjacent pieces can be lifted off with the edge of a finger nail or by masking tape.

There is a standard test for adhesion but it is destructive. A multi-bladed cutter is dragged across the surface of the paint to create some parallel lines. Then the cutter is drawn across these lines at right angles to produce a crossword grid. How the paint breaks away from all these sharp corners determines how good the adhesion is. It is not normally used in the bodyshop but it is the only way to resolve a dispute which has reached an impasse. Several car manufacturers have a device that fires shot at a small square of paint at a pre-determined pressure. The results are compared to a series of photographs to determine the paint condition.

Rectifying defects

Remedial action

This element consists of advice which you may pass on to vehicle users to help them avoid the consequences of abuse or poor care, and some pointers to help in deciding how to rectify defective paintwork.

The best advice to users is to avoid the circumstance that causes the defect in the first place. This is not always possible, of course. Acid rain will affect a car wherever it is in the open. Avoiding parking under trees will prevent aphid or resin damage and even minimise that from bird droppings. Advising a customer to park away from a railway line may well prevent ferrous particle contamination.

For all environmental and industrial fall-out problems the best preventive treatment is a good coat of wax. Advise the customer that this can be checked whenever it rains; if the water collects into globules, the surface has a coating of wax. A heavy coating of wax is essential for all vehicles that must be parked for long periods beneath trees or beside busy railway lines.

Stone chipping damage to sound paintwork can be reduced or even largely eliminated by not driving too close to the vehicle in front or by slowing down on loose surfaces. For those who use rough tracks or drives it may be beneficial to apply a transparent plastic film to the more vulnerable paintwork.

Where rectification is involved there are, broadly speaking, three main methods:

■ The paint film is sound but has surface blemishes which are only in the topcoat; the defect can usually be removed and the gloss restored by polishing or flatting and polishing.
■ The paint film is sound but there are defects showing through from below such as sanding marks or substrate contamination. This calls for flatting down or stripping back the paint, making good the defect and re-finishing.
■ Where the paint film is defective, as is the case with soft paint or poor adhesion, the paint must be removed as far as necessary and the vehicle or affected areas refinished.

The correct diagnosis is therefore crucial in deciding whether the work will take a few minutes or become a major refinish job.

Materials

Additional materials that are needed for defect rectification range from sheet and liquid abrasives to acid. Paint makers provide special products such as fade-out thinners for spot or local repairs, as well as advising on techniques for sealing existing paint coatings. Always follow the guidance of the paint company whose products that you use when over-coating existing finishes and when carrying out local repairs.

The abrasive papers usually used to level out a paint film surface range from P1200 to P2000. Specially cut small discs of a suitable grade are also available for the purpose of small defect removal such as de-nibbing.

Polishing compounds are available for all types of finish improvement or for restoring gloss after flatting with abrasives. Some makers offer a choice of compounds and matching compounding heads to enable a good finish to be obtained on any basically sound paintwork, whatever the defect. As with paint systems, it is advisable to use only one scheme and follow it through to completion on a rectification job.

Where industrial pollution is a problem there are a couple of methods which work well in most circumstances. A light marking of sooty spots on a pale coloured finish can often be removed with a non-abrasive household cleanser which contains a little mild bleaching agent. A typical agent in these cleansers is ammonia. However, you should be aware that ammonia can damage acrylic resins if it remains on the surface so any residue must be thoroughly washed away.

Ferrous particle contamination will require some dilute oxalic acid, which is obtainable from vehicle manufacturers or industrial chemists. A small, gentle brush is needed for the more stubborn deposits.

Rectifying paint surface defects

A paint defect is a blemish which the viewer considers should not be there. In the United Kingdom a relatively high standard of finish has come to be accepted as the norm,

despite the difficult conditions that some painters experience. Most of us expect the panels on the more visible parts of a vehicle, the bonnet, upper doors, roof edges and boot lid or tailgate, to be free of noticeable blemishes. Tight cost constraints mean that whenever possible the finish should be acceptable 'from-the-gun'. If there is some rectification to do, then it must be done easily if there is to be any profit made from the job. That calls for a quick, effective remedy. It is even better if the problem is noticed during painting and dealt with then. Inclusions, particularly the fibrous type, can often be lifted off a wet paint film with a small screw of masking tape used sticky side out, allowing the paint to flow over the mark. This is particularly useful for blemishes in basecoats. Severe contamination on fresh paint is best removed while still wet by wiping off the whole panel and cleaning with fresh wipes and degreaser.

One of the most common surface defects, inclusions, will need to be remedied by every painter from time to time, however good they are. Individual inclusions are best removed with a de-nibbing block and a very fine grade of abrasive such as P1700 (see Fig. 141). The use of a small, firm block will enable the rubbing to concentrate on the protrusion and minimise the abrading of the surrounding paint film. Once the nib has been flatted back, the gloss can be restored by polishing or buffing. Be guided by the maker's recommendations for the system you are using.

A pneumatic or an electric polisher is the easiest and quickest way of restoring the gloss. A special compounding head should be used, often available from the compound maker. This is the process using a Farécla head and compound, step by step:

■ The head is soaked in water before use, squeezed out and fitted to the machine. Water acts as an essential lubricant with the compound, enabling it to cut away the paint and, at the same time, helping to keep the panel cool.
■ A little compound is spread on the head. Avoid using too much; it simply flies off and must be cleaned up afterwards.
■ Close the open compound container. This is vitally important as just one piece of grit in the compound can change a minor job to a major one.
■ Lower the mop-head onto the panel and start compounding. The speed should be under 2000 rpm and the head should be held at a slight angle to the panel.
■ The polisher is gently moved all the time to avoid heat build-up.
■ Use a water sprayer to add moisture if there is any tendency for the surface to dry out.
■ As the cutting action falls away just add a little water and keep the mop moving. Allow the face of the mop to polish the surface and ease the weight of the machine to improve the gloss.
■ Polish off with a soft, open weave cloth to check the surface. Repeat the process if necessary.

Another famous make, 3M®, uses an almost identical process except that the mop is initially soaked in the compound because water is not used as a lubricant. Whichever system is used in your bodyshop, always follow the maker's directions.

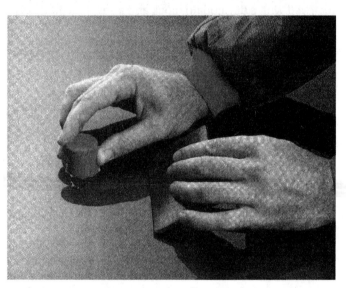

Figure 141 Isolated inclusions are best removed with a very fine abrasive on a small block

Another common defect is the paint run. If it appears to be cured the surface is cut open with either a special 'plane' or a piece of body file. If the paint inside is still soft it must be cured, the easiest method being a few minutes with a hand-held infra-red heater. When all the paint is fully cured, the excess must be cut back as far as possible with the cutter. The remains of the run are then removed with 1200 grade wet-or-dry abrasive. The gloss can then be restored using the technique already described.

Polishing techniques can be used to overcome other defects, such as poor gloss, providing the paint film itself is sound. Always bear in mind however that the finish should be as close as possible to the standard of the remaining panels. A slight although perfectly acceptable orange peel factory finish can be made to look decidedly shabby beside a repainted, mirror finish bonnet.

Ferrous particle contamination is removed by washing the affected panels, invariably those facing upwards, with a 10:1 mix of water and oxalic acid. This must be done on a designated washbay and the operator must be fully pro-

tected against acid splash. In certain circumstances, usually of poor ventilation, it is possible to be affected by the fumes. If there is any risk of this occurring a ventilated visor should be worn.

If the contamination is severe, as is likely to be the case with a vehicle which has not been waxed regularly, a brush will be needed. The bristles should be stiff enough to dislodge the particles but not so stiff as to scratch the paint. A medium texture tooth brush, though small, is ideal. It also has the advantage that the plastic handle can be given a chisel shape for dislodging stubborn particles. This should be used in conjunction with a magnifying glass.

Once all contamination has been removed the vehicle should be thoroughly washed, preferably with a high pressure washer, to remove all traces of the acid. Particular attention must be given to rubbers or any places where acid may be lodged. When dry, the vehicle should be given a heavy coat of wax to minimise any further contamination.

Rectifying paint and substrate defects

We are concerned here with some aspects of refinishing rather than with surface rectification and this will usually involve at least some of the paint being removed. This will be necessary where the paint depth is too great to accept further coats or where the paint coating is defective. Where there is no reason to suspect the foundation coats, only the topcoat need be sanded back. An overhard filler coat or a poorly applied or finished filler coat may need more substantial preparation. If the defect stems from too much paint altogether then the finish must be taken back to an acceptable depth for the refinish coats. The topic of paint depth is dealt with in Chapter 6.

The method to be used for removing the finish for repainting depends upon the finish, the level to which refinishing must go, and the source of the defect. Where the whole coating is defective or there is a problem with the substrate then the paint must be removed in total. If all of the paint on some or all of the panels must be removed it may be more suitable to use a paint stripper. For a local problem, such as a small area of substrate contamination, sanding down is the most effective method.

In all other respects follow the advice and use the materials of your paint maker in the manner given in their data sheets.

Chapter 9 | Augmentation – Personalising Vehicles

Advising Customers on Modifications

Warranty and legal implications

The extensive modification of vehicles is not an activity in which many bodyshops engage. It is also an activity which does not generally meet with the approval of vehicle makers if fundamental changes to the design are envisaged, whether it be to the mechanical components, the systems or the bodywork. Only accessories or modifications which are approved by the vehicle maker may be added or made during the warranty period if the customer wishes to retain its benefits. It may also be necessary for these to be fitted or carried out by an approved dealer or bodyshop.

With some body warranties providing cover for ten years, the situation must be clearly explained and no work undertaken until formal consent and release from the liabilities of the warranty have been given by the owner. Given the tendency of the courts to lean in the direction of the consumer and the uncertainty of legislation in this area, any modifications must be a matter for serious consideration and such contracts should never be entered into lightly.

Where a vehicle is no longer under any warranty, the main concerns will be the safety of the modified vehicle, its compliance with the Construction and Use Regulations and any other requirements of the Road Traffic Acts. The simple job of adding some low-slung fog lamps, for example, will dictate how they are wired into the vehicle's electrical circuits.

Subject to these conditions, however, there are additions and restyling modifications which may be applied and which may not affect the warranties, the safety of the vehicle or contravene the Construction and Use Regulations.

Information sources

The first information source must be the vehicle maker to see if there are any approved modifications or fitments available that would meet the customer's needs. Enquiries can also be made as to who may carry out the work and what specifications exist for it.

Ready made components and accessories are normally supplied with instructions and, where these are approved by a vehicle maker, will not normally invalidate any warranty. Where any work is carried out on the body, however, even if it is only the drilling of some screw or rivet holes, every care must be taken to see that special coatings or anti-corrosion measures are carried out as required by the vehicle maker.

Features, benefits, disadvantages and costs

A customer who asks for a quotation may be doing so merely out of curiosity or they may already know a considerable amount about the proposed work. Some nice judgement is called for if you are to satisfy curiosity and gain a customer. There is no point in spending a lot of time on researching a project which is not likely to be undertaken. However it is important to supply sufficient information to gain a worthwhile customer. The serious enquirer should be treated seriously, and only carefully considered information should be supplied as this may well form the basis of a contract to carry out the work.

It is in the interest of both the bodyshop and the customer if the features of the proposed modification or accessory are reviewed together and discussed. Motorists tend to gain some peculiar ideas about the value of changes and additions, sometimes encouraged by promotional advertising. The addition of rear spoilers is an instance. These may have little or no effect at speeds below the legal limit of 70 mph and so their ability to add to the 'road hugging' capabilities of the car will be of little or no practical value; the money would be better spent on servicing the suspension or on better tyres.

If the benefit that the customer seeks from the rear spoiler is to alter the appearance of an otherwise standard car to satisfy themselves and impress others, then they will be happy with a job that looks good. Under no circumstances should they be led to believe that cornering at legal speeds will be in any way improved.

Personalise the Vehicle to Specification

Health and safety at work and PPE

There are no particular requirements so far as personalising is concerned, and general working precautions have already been covered in the earlier chapters. For information on PPE and other measures when fitting components or carrying out panel work refer to Chapters 3 and 4; for advice on health and safety matters relating to refinishing refer to Chapters 6 and 7.

Job	Source of information
Fitting accessories	Vehicle makers' instructions Accessory makers' instructions Specialist information, e.g. sealer makers' data
Bodywork modifications	Specialist publications
For information on how the vehicle is constructed and any special considerations e.g. anti-corrosion measures see:	Vehicle makers' information MIRRC (Thatcham) information Proprietary manuals
Special effects painting	Paint makers' information Specialist publications Equipment makers' information

Information sources

Bodyshops which carry out this type of work may already have at least one member of staff who is familiar with the trends and practices currently in vogue. The skills involved are often quite specialised and incorporate many of the ideas of the individual craftsperson. They are probably the best source of information on the subject (see Figs 142–144).

Figure 142 Applying a coachline from a tape

Equipment and tooling

All the usual equipment and tooling may be needed, depending upon the job, of course. There are some variations and here are two of them.

The fitting of sunroofs is a profitable job for the bodyshop, rewarding for the installer and generally of great benefit to the user. Few tools are needed and amongst these many operatives favour a jigsaw for the purpose of cutting the roof opening. It should have a metal cutting blade, of course. Alternatively, an oscillating saw or cutter may be used. Because of the small particles created by these saws the remainder of the vehicle must be covered to prevent them gathering in recesses or lodging under flexible trim. Large,

reusable dust sheets are probably the best protection. Some car makers prefer the use of a nibbler, as it reduces the possibility of swarf becoming trapped inside the roof cavity and causing corrosion.

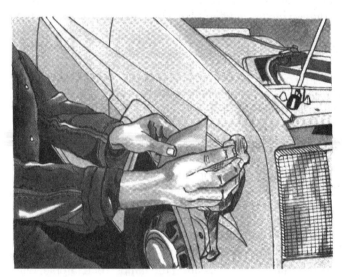

Figure 143 Applying decals or decor

Figure 144 Air must be squeezed from wide decals as they are applied to avoid bubbles

Special effect painting will probably call for the use of the small spraygun called an airbrush. This permits fine details and other effects to be created.

Panelwork modifications and fitments

Such work ranges from adding pre-prepared components, such as wing extensions, to that of making totally new components involving the traditional skills of the body-builder. It is not within the scope of this book to cover such an involved and very practical subject. There are, however, some important considerations which should be borne in mind if any work of this nature is undertaken.

Changing the original shape of a vehicle may have far-reaching consequences on both the safety and correct operation of components. Alterations to the wheel arches, for example, should take into account the full suspension movement when the vehicle is fitted with the intended wheels. It also follows that the lock-to-lock movement of the wheels should be checked. In both these aspects provision must be made for the existing movements or limiters provided to accommodate the new design. Failure to do so could result in an accident from tyre blow-out.

Where wheels are to be changed as part of a re-design, the safety features built into the suspension and steering should also be considered. For example, Volkswagen has used 'negative roll radius' on most of its models until recently, and is continuing to do so during the production life of existing models. This design feature provides what is known as sta-bilised steering. Where this is employed on a car the result of uneven braking or even a flat tyre on the front wheels causes no inconvenience (other than the need to change the wheel, of course). The driver retains perfect control and there is no fear of running off the road. Fitting a wheel with a different offset will alter this geometry and, as it is only a matter of a few millimetres, can easily remove it altogether.

One other point concerns the cooling of the brakes. Motorists have complained for many years about the way brake dust gathers on the outside of roadwheels on some vehicles. It collects there because the rotating wheel draws air over the brake to cool it and in doing so carries the dust to the outside. Changing the wheels may easily overcome this problem. There is every chance though that loss of the brakes may occur due to brake fade.

Providing that wheel changes are made with the custo-mer's consent and with the realisation that safety features may have been lost, there may be no cause for concern. If, however, the customer is not advised of the risk and the work is done, your bodyshop or even you may be held liable in the event of a subsequent accident.

Many vehicles today have sophisticated suspension sys-tems and a modest change to a wheel size may bring about significant changes to suspension geometry and the car's handling. To avoid such pitfalls only ever replace wheels with those supplied or at least approved by the vehicle maker.

Another important consideration is the effect of changes to the airflow, not only to the way the vehicle handles but also to the cooling of the engine, the transmission and the fuel system. Minor changes to the outline shape can sub-stantially alter the relative flow of air over the top of the body and underneath. This can decide whether the vehicle will be lifted or pressed down towards the road. The air flow around and even below the body may also have an effect upon the cooling of the engine compartment. Whilst the cooling fan may be able to cope with a modest increase in temperature, the temperature of the fuel lines could rise to the vaporisa-tion point, despite them having a circulating system.

Special effect painting

Some of the major paint makers provide materials and infor-mation specifically for painting in this way (see Fig. 145). Training courses are offered in the necessary skills, enabling the painter to develop the creative element so essential in this type of work. Any painter wishing to become successful in special effect painting must first become proficient in the skills needed for normal refinishing.

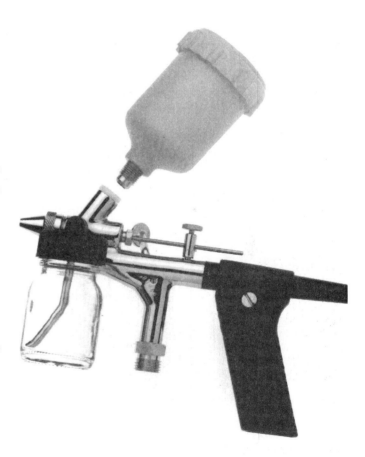

Figure 145 The Sata Dekor Z for customised paintwork design

Chapter 10

Effective Working Relationships

Customer Relations

Employees in any company have jobs because there is work to be done; that work is there because customers are buying its goods or services. Without customers, any business collapses and its workers are made redundant. Customers are not interruptions to your routine of work; they provide it, and in so doing ultimately provide your wages.

Customer requirements and expectations

We are all customers, buying the things that we need to live and, perhaps, some extras that give us pleasure. Each of us measures up the goods or services we buy and decide whether they have given us value for our money. If a cheap pair of shoes wears out much too quickly, we will probably think that that is not surprising and maybe we do not buy any more of that type. Expensive shoes that wear out quickly are a different matter; the shop soon knows about it and we insist on the matter being put right!

The customers who come to the bodyshop where you work are no different, except that they, and probably their insurers, will be spending rather more money than the cost of a pair of shoes. The situation is similar in another way, too. Most of us buy work shoes only when we need them – it's what might be called a 'distress purchase'. Customers who come to your bodyshop are, in the main, not there by choice either. They have, through their own fault or that of someone else, damaged their car and need to have it repaired. That is very much a 'distress purchase'. Worst of all, they have lost their means of transport or even, sometimes, the means whereby they earn a living.

Importance of the customer

If we and the company in which we work are to thrive, the customer must come first, always. A customer who feels that you have been listening, who thinks you have done the best that you could for them, that you always seemed to want to help them and that you have done a good job, will be the best advertisement that it is possible to have.

The opposite of that is even more effective in the worst possible way. A customer who is unhappy and totally dissatisfied will tell all and sundry at the slightest provocation. It is quite natural for us to brood on what we see as misfortune and to share it with others.

You must care for your customer, then, even though they may not come back they will, hopefully, tell others of your good work. Your reputation will be built on the care that is given to each and every aspect of the job. You may never meet, but the work that has been done well in good time clearly shows that you do care. If that is true you can be certain that the word will get around.

Customer relations techniques

Taking care of your customer does not mean that you must pander to their every whim and wish. What is important is to convince them that you have their best interests at heart and that you will use every endeavour to meet the promises that you have made.

We all, as customers, expect everything to be done at once, to perfection. Sometimes a 'miracle' does happen, but we are usually brought face to face with reality regarding cost, delivery, or even the possibility of doing it at all. How that is done often decides whether we will become a customer, or how we tell others about the company.

Many bodyshop customers are captive in a sense because the damaged vehicle has been recovered there, or because the company is an approved repairer of the insurer. The customer does have the right, however, to have the vehicle repaired elsewhere and may do so to comply with the body warranty or if the right relationship is not set up. It must also be remembered that many will not be hardened to having accidents, even if there has been no injury, and they may well be in some distress when they make contact, especially if they arrive with the damaged vehicle. They will remember kindness and concern shown to them at this time and appreciate it.

To establish the right relationship demands that you 'qualify your customer', a sales term that simply means deciding what you need to know about them. Remembering that they may behave a little irrationally, a great deal can be discovered simply by observing their body language. We all recognise someone very like ourselves and know how to behave towards them. Nonetheless, a friendly respect towards the customer will probably be appreciated.

If the customer's authoritative appearance and manner suggest someone who is used to making decisions and is normally in control, recognise the fact by talking to them politely, by being positive and by conversing at their level. Such people will grasp facts very quickly and will not need repetition or detailed explanation. If they do, they will demand it.

The busy lady customer, just like many men, will not necessarily know very much at all of the procedure, and much less of the technical terms that everyone is using. Watch for signs of a vacant expression, one of the best ways of knowing if a person understands what you are saying. A friendly query will often win an honest response and you can then build a relationship.

Never make easy promises which have to be explained away later when things go wrong. Be as reasonably honest as you can without frightening the customer. Here again, watch for reactions and adjust what you say to them.

Your own body language, as well as what you say, is vitally important. Avoid all those unhappy experiences that you have probably been subjected to as a customer. Always give a customer your full attention and listen to what they have to say. If possible go to an area where there is not too much distraction so that you can both concentrate. Both of you should sit down, preferably on the same side of a table. A table is necessary for the inevitable forms or notepad, though you should never let it come between you as a barrier. Listening is a skill and needs concentration. If you cannot understand what the customer is saying, ask questions to find out. Try to understand the concerns the customer may have and take any reasonable steps within your power to help, perhaps by seeking guidance from your manager or supervisor.

Customer care is as much a skill for which training can be given as any other activity. If you have the opportunity of a training course take it; learn the skills and practice them.

Dealing with complaints

There are plenty of opportunities in our industry for things to go wrong and, of course, they sometimes do. Dealing with them involves everything that we have discussed in the previous topic, plus a lot of tact and honesty.

Never try to get out of a problem by lying about the situation, either on your own account or to protect someone else. Very often the truth becomes known later and the result is at least embarrassment or possibly litigation. Invariably, the company policy will ask you to be polite, to tell the customer that you must inform your supervisor or manager immediately of the problem, and do so.

Security of customers' property

All bodyshops should have facilities for the safe keeping of customers' property. When attending a recovery or dealing with a new job any loose or easily removed equipment or possessions must be gathered together, listed if the company policy asks for that and bagged for careful storage.

The vehicle itself is in your hands and it is evidence of your care that the correct protection is applied and any other measures are carried out as needed. Components removed from it should be stored safely in a designated rack and clearly labelled so that they are refitted again.

Working Relationships

Lines of communication

All working groups have methods of communicating information about their work and this is especially important in the bodyshop, whether it is large or small. The difference will be the degree of formality and the person to whom a report is made. Whatever the policy in your company, as soon as there is any information to pass on, it must be done promptly and in the correct way.

Information relating to the safe operation of equipment, the provision of protective equipment and the notification of potential hazards such as noise are all areas of routine which must be brought to the attention of a supervisor or the manager.

This is also an area where caring for the customer can be seen in action. The moment that a variation from an estimate, or a damaged or missing part is noticed, it must be communicated to those who can do something about it to avoid unnecessary delay.

Levels of authority

In a very small bodyshop, the staff may report directly to the proprietor. In a medium size operation there could be a supervisor or manager who handles day-to-day queries and job progress.

Much larger bodyshops are likely to have administrative personnel to whom routine reports about the work are made. Technical queries will probably be dealt with in the first instance by a chargehand or team leader. In such an organisation the lines of communication will be well defined and laid out in the company operating procedures.

Interdependence of all staff

The success of a company certainly depends upon how it looks after its customers. However, whilst that is very important, the way that the staff work together also plays a critical role in any success. It is probably of greater importance in the smaller bodyshop where, for example, much of the communicating is done by word of mouth and relies on the relationship of trust between the individual members of staff.

Relationships sometimes gel together without any apparent effort on the part of the people concerned and any such group is fortunate. That is not always the case and as a member of any group you must always be prepared to put some effort into creating a working relationship between yourself and the others involved. The basis of trust at the root of such relationships nearly always follows an indication of genuine care and concern.

No great demonstration or undue attention is needed. A little thought during work about how others will be affected by what you do will show quite clearly that you want to be part of the team. The old adage is still true: 'do as you would be done by!'

Company image

This is a topic that is not often discussed and yet is an important aspect for many. Most people like to feel that they are part of a team and want to feel pride in their work. That is as it should be and makes the working day more meaningful than just existing between clocking in and clocking off. Caring for the customer and working together as a team are all part of this equation. The other element is caring about how the bodyshop appears to the outside world and to those who come into it.

Your daily work rapidly becomes routine and routine has a habit of gradually adopting some of the actions that show others a less desirable image. Personal appearance is important. Yes, of course, working in the panel shop or fitting bay can be very dirty. Your hair may well look like a mop sometimes. If you are visible to customers or visitors, brush excess dust off your overalls occasionally and perhaps tidy your hair. Regardless of whether you can be seen, your overalls should be worn as intended and not tied around your waist.

The safety benefits of keeping the workbay as tidy as possible have already been mentioned. To this should be added that the simple action of dropping waste wrapping or wipes on the floor rather than into a bin displays an uncaring attitude. Shouting across the workplace in the presence of visitors does nothing for the company's image.

Remember, too, when discussing your workplace with mates after hours that others may hear a snatch of conversation and place a wrong interpretation on what is said. You are, however unwittingly, an ambassador of your bodyshop when others hear that you work for them.

You have worked hard to learn your skills and, no doubt, want to improve them if you can. Have pride in your bodyshop. Most importantly, have pride in yourself and the skills of your craft (see Fig. 146).

Figure 146 Considerate handover of a good job is the final proof of caring for the customer

Index